D0713495

Designing Public Procurement Policy in Developing Countries

Murat A. Yülek • Travis K. Taylor
Editors

Designing Public Procurement Policy in Developing Countries

How to Foster Technology Transfer and Industrialization in the Global Economy

Editors
Murat A. Yülek, Ph.D.
THK University
Ankara, Turkey
myulek@pglobal.com.tr
mayulek@thk.edu.tr

Travis K. Taylor, Ph.D.
Christopher Newport University
1 University Place
Newport News, VA 23606, USA
ttaylor@cnu.edu

ISBN 978-1-4614-1441-4 e-ISBN 978-1-4614-1442-1
DOI 10.1007/978-1-4614-1442-1
Springer New York Dordrecht Heidelberg London

Library of Congress Control Number: 2011941385

Printed on acid-free paper

Springer is part of Springer Science+Business Media (www.springer.com)

To my parents, wife, and children
Murat A. Yülek

To my wife Sunny, and daughter Nari, for your love and encouragement
Travis K. Taylor

About the Editors

Murat A. Yülek, Ph.D. Murat A. Yülek is the Vice Rector and Dean of Business Faculty at THK University, Ankara, Turkey. He has authored a number of books and articles in economics, development, and finance. He is the Chairman of the Board of Directors at an investment bank and a partner at an economic consultancy firm. Previously, he had positions at the International Monetary Fund and taught at Georgetown and Bilkent Universities. He holds Ph.D. and M.A. degrees from Bilkent University, MBA degree from Yale University, MSM (Management) degree from Boston University, and a B.Sc. degree from Bogazici University.

Travis K. Taylor, Ph.D. Travis K. Taylor is an Associate Professor of Economics at Christopher Newport University, Virginia, USA. Previously, he served as Chair of the Department of Accounting, Economics & Finance at Christopher Newport University, and International Economist with the US Government. Dr. Taylor has been a visiting professor at the University of Richmond, and a visiting scholar at the Australian Defence Force Academy in Canberra, Australia. His research interests include contracting and economic development, and he has published a number of articles on the subject of international procurement. He holds Ph.D. and M.A. degrees from the University of Connecticut, and a B.A. from the University of Richmond.

Preface

In October 2010, PGlobal Ltd and Istanbul Commerce University hosted a workshop on *Designing Efficient Policies to Foster Technology Transfer and Development Capacity in Emerging Markets* in Istanbul, Turkey. The workshop brought together scholars, policymakers, and industry specialists from the fields of development, technology transfer, and procurement contracting. The contributions to this volume were edited papers from the workshop, which explore the feasibility of procurement policy to influence a country's industrialization, technology, and trade (ITT) trajectories.

Economies grow by increasing inputs (e.g., capital and labor) and productivity. Productivity or, in the parlance of growth accounting, *extensive growth* derives from improved technology, institutions, and other efficiency enhancements. Each country must grapple with competing philosophies and policy proposals to foster an environment that is conducive to technology acquisition and absorption.

Economists have long debated the role of government in economic activity and growth. Noting the existence of imperfect information, transaction costs, and increasing returns, some argue in favor of a planned economic development strategy that usually includes industrial policy. On the other end of the theoretical spectrum is free-market capitalism, which rejects any government intervention on the grounds that such actions are economically inefficient. Of course, most development strategy is far more nuanced, and one is hard pressed to find a country that is entirely centrally planned or *laissez-faire*.

Four theories of economic development have dominated the economic development literature since World War II. The linear stages model was most prominent during the 1950s and 1960s. Worldwide recession and stagflation in the 1970s spurred new and competing theories from the structuralist and dependency schools. The rise of neoclassical economics spawned renewed interest in free-market capitalism and private property rights in the 1980s. Government corruption, bureaucratic incompetence, and meager growth in many developing countries also contributed to the resurgence of laissez-faire economic policy. At the same time, neoclassical research and policy studies culminated in the so-called Washington Consensus policy program in the 1990s.[1]

In recent years, however, a call for a more tempered, moderate view of development policy has emerged. Proponents of the New Consensus note that some of the most robust and equitable growth has occurred in countries—most notably in East Asia—that deviated significantly from the policy prescriptions of the Washington Consensus.[2] The more moderate view gaining advocates among development economists and policymakers retains the central tenets of private sector competition, fiscal and monetary prudence, and limited government.[3] However, this view diverges from the Washington Consensus by acknowledging a planning role for developing country governments to address market failures (and missing markets), establish institutions that support long-term growth, and implement policies that influence ITT. The moderate view, then, is the foundation upon which this volume stands.

Brief Overview of Industrial Policy and the Public Procurement Link

In perfectly competitive market environments, most economists agree that there is little role for government intervention.[4] Developing countries, however, are rife with market imperfections, externalities, and poor infrastructural investment. Public goods, such as defense, roads, and utility lines, are frequently undersupplied. As a result, a case for targeted government intervention in developing countries is not inconsistent with the standard neoclassical economic model. Indeed, the theory of second best holds that when multiple markets in an economy fail to meet the stringent requirements for perfect competition it may be welfare increasing to employ a gradualist approach to international economic integration (see Lipsey and Lancaster (1956).

Industrial policy utilizes targeted government intervention to augment markets and promote the production of goods and services that are considered in that country's strategic interest. The argument that governments somehow have better information than private firms about the prospects of a firm or industry—and is thus worth targeting—is not persuasive to many economists. Furthermore, if a firm requires subsidies or protection to exploit the gains derived from learning-by-doing and scale economies, profit-seeking capital market participants should be able to provide the necessary financing. Similar arguments invoking perfect

[1] For an overview of the Washington Consensus, see Williamson (1990).

[2] The World Bank (1993) report illustrates how the East Asian tiger countries employed a successful policy mix that balanced market competition and government support for targeted industries.

[3] Rodrik (2008) offers a critical assessment and juxtaposition of the policy prescriptions.

[4] Even the most ardent free-market supporters usually accept limited roles for government including the provision of public goods and national security, enforcement of contracts, and central banking.

information and efficient capital markets are frequently put forth to discredit industrial policy in the presence of coordination failures, spillovers, externalities, and the appropriation of firm research and development (R&D). Practical experience of numerous countries, however, has shown that such policies can foster growth and accelerate economic development.[5]

In countries with competitive markets, the theoretical case for procurement and industrial policy diminishes significantly. Grossman (1989) reviews the literature and refutes most of the arguments that support industrial policy.[6] Pack and Saggi (2006, p. 267) arrive at a similar conclusion, although they maintain that "market failures can, in principle, justify the use of industrial policy." Instead, the authors argue for a shift of focus away from industrial policy, per se to "negotiation with multinational firms on issues ranging from environmental regulation and taxes to efforts ensuring local learning" (293).

The efficient market structures and institutions that obdurate the need for government involvement, however, do not typically exist in developing countries. In recent years, contributions to the literature from Chang (2008) and Reinert (2008) have highlighted the importance of industrial policy for all countries as they graduate through the stages of development. Nobel laureate Michael Spence (2011) argues that even the industrialized economies of today ought to follow the recent path of Germany, which has carefully supported select high value-added industries. And while the theoretical debate on the economic efficiency of industrial policy rages on, most governments around the world support—in practice—a limited role for government to foster critical industries.[7]

Procurement Policy as a Form of Industrial Policy

The discussion of government policy as a means to achieve the ends of technology acquisition and industrialization is not, then, unique to this book. Rather, the focus here is on procurement policy as a lesser-known instrument of industrial policy. A voluminous literature on industrial policy exists, and the number and quality of procurement policy (domestic and international) studies have increased significantly during the last 20 years. However, analysis of procurement policy as a strategic instrument of industrial policy has received relatively less attention from scholars (see, e.g., Li (2011); Kattel and Veiko (2010); Eliasson (2010); Edler and Georghui (2007); Bolton (2006); Uryu (2006); Watermeyer (2000); and Geroski (1990)).

---

[5] Some East Asian country policies are, perhaps, the most prominent and recent example. However, at similar stages in their development countries such as England, the United States, and France made extensive use of industrial and trade policy.

[6] Grossman's analysis draws heavily on Baldwin (1969).

[7] Special financing terms, tax exemptions, and other incentives can also be extended to select firms for the purpose of creating "national champions" (e.g., Siemens in Germany; Nokia in Finland).

Several chapters included in this volume make the case for well-conceived industrial targeting via public procurement policy. As one of the largest buyers in many product markets, governments have considerable leverage to influence the terms of a transaction. Negotiating for price discounts is the most common and recognizable way governments can benefit. However, a menu of other policies—that may or may not be preferable to price discounts—is available to procurement officials as well. These policies include, but are not limited to[8]:

- Preferential vendor and/or industry purchasing arrangements
- Domestic preference
- Local content
- Countertrade and offsets

Preferential procurement terms granted to select domestic firm(s) enable a government to promote a particular industry in what amounts to a production subsidy. Such action will invariably alter the allocation of resources, prices, and welfare in the economy. Therefore, this policy should be adopted only after a careful analysis of costs and benefits. Government support for the so-called "green technology" falls in this category; the seemingly banal SWaM (small, women and minorities) preference given in American states does as well. In the case of the latter, for example, it is widely known that a product procured from a SWaM vendor may carry a higher price tag compared to a competitor's offering of equal quality. In most product markets, governments enjoy numerous purchasing options in the global economy. Open competition—both domestic and international—is generally preferable in procurement because the price and quality competition leads to increased welfare in the purchasing government's economy. Moreover, in the absence of explicit or implicit protection from the home government, domestic firms are compelled to become more efficient and innovative.

However, it may be in a country's interest to extend implicit (subsidy) or explicit ("buy domestic") terms to domestic firms in select instances. In 1933, the United States, for example, passed the Buy America Act, which requires the federal government to "buy domestic articles, materials, and supplies when they are acquired for public use unless a specific exemption applies" (Luckey 2009, p. 5). Numerous exceptions to the policy exist, but the substantive aspects affecting strategic industries and the general intent remain.[9] The World Trade Organization's General Procurement Agreement (GPA) seeks to eliminate or at least limit such protectionist policies among members, but exemptions exist for products deemed of

[8] Another example is "planned and pre-announced" public procurement, which requires sufficient government credibility to induce capital expenditures and R&D investments in the private sector. This policy also requires an effective and well-planned government. In many developing countries neither of these conditions is met. Therefore, planned procurement is probably better suited for developed economies with efficient institutions.

[9] Exceptions to the Buy America Act fall under five categories: (transaction) inconsistency with the public interest, unreasonable cost, products that will be used outside the country, products produced in the United States of insufficient supply or quality, and procurements under $2,500.

national security interest, public health, and the environment. Developing country members are also exempt, unless they voluntarily sign the plurilateral agreement.[10]

Local content rules require a percentage of the procurement workload to be fulfilled by firms in the domestic economy. This form of intervention may help a developing economy in two ways. First, local content rules generate additional work orders for domestic businesses. When a government procures a product from a foreign supplier, the domestic multiplier effects are negligible. When at least part of the procurement calls for local content production, the domestic economy benefits from multiplier effects via increased economic activity, employment, and income. Second, local content rules help domestic businesses acquire know-how, transfer technology, and lower unit costs through learning curve effects.

Countertrade agreements represent another class of international procurement policy instruments. Countertrade contracts can be designed to alter the ITT mix of a country. These contracts build reciprocity into the transaction by requiring a foreign seller to purchase specified products from domestic firms. Barter, counterpurchase, buyback, and offsets are the most common examples. And while these contracts vary from one another, two commonalities exist: (1) departure from the price margin of arm's-length exchange and (2) conditionality.

Under a countertrade procurement contract, a government may elect to use its oligopsony power in one market to bargain for reciprocal contracts that—in lieu of price discounts—shift the terms of trade off the price margin. Sometimes referred to as "non-standard," the contract is hardly uncommon. Most estimates put countertrade at 15–20% of the total world trade. Although the bilateral (conditional) nature of countertrade contracts can induce trade diversion and the associated world welfare effects, the purchasing government's economy may benefit from foreign exchange savings, increased work orders, export market penetration, technology transfer, learning by doing, and reputational economies.

Counterpurchase agreements may reap similar benefits, although hard currency savings have been shown not to be a motivating factor since cash still changes hands. Historically, countertrade contracts have been employed most often by natural resource-abundant developing countries. For example, the Malaysians have bartered palm oil, the Russians oil and natural gas, and the Thai Government struck an agreement to pay foreign firms with chicken wings. In a world of imperfect information and transaction costs, countertrade contracts can be crafted to open new markets and establish relationships with foreign firms, thereby raising welfare in a manner described by the theory of the second best.

Offset arrangements require the foreign firm to transfer economic benefits (beyond cash) to the purchasing government's economy as a condition for the sale of the base product. These benefits may include the aforementioned countertrade instruments or a myriad of other arrangements that vary in complexity and time to fulfillment. Technology transfer, managerial services, investment, credit transfer, licensed production, co-production, and loan-import agreements are some

[10] At the time of writing, 40 of the 153 WTO members were signatories to the GPA.

of the more common vehicles to transfer the benefits. Procurement data reveal that while simple countertrade contracts are historically favored by less-developed countries, offsets are the instrument of choice for middle- and high-income countries. These preferences can be explained by noting that the relatively sophisticated offset arrangements are designed to achieve multiple development objectives. In this respect, public procurement is a platform from which a set of policies can be crafted in support of an overarching economic development strategy.

The Chapters

The book is divided into three parts: (1) theory and policy of procurement as a tool to foster technology transfer and industrialization, (2) country experiences, and (3) case studies of particular industries.

The five chapters comprising Part I focus, collectively, on procurement policy as a critical part of a comprehensive development strategy. In Chap.1, Murat Yulek examines the growth of public expenditures in developing countries and forecasts public sector purchases of machinery and equipment. His analysis sheds light on the oligopsony power of many purchasing governments, and the largely untapped potential to extract rents from multinational corporations. An important theme of the entire book is introduced: namely, a well-conceived procurement policy can affect the rate of technology acquisition and industrialization.

This thesis is demonstrated using theory, policy application, and empirical data in Chaps. 2 through 5. Travis Taylor examines the empirical record of offset arrangements in international government procurement (Chap. 2), and Ron Watermeyer develops a framework for governments to link procurement to development outcomes (Chap. 3). Houssam-Eddine Bessam, Rainer Gadow, and Ulli Arnold revisit import substituting industrialization policy and argue that it still has a place in the developing-country toolkit (Chap. 4). Mahmut Kiper explores the relationship between knowledge, technology transfer, and economic development (Chap. 5). Attention is given to the challenges developing countries face in obtaining and absorbing technologies that help create comparative advantage.

Parts II and III present selected country experiences and industry case studies. Analyses of South Africa (Chap. 6), South Korea (Chap. 7), China (Chap. 8), and Turkey (Chaps. 9 and 10) offer firsthand accounts and micro-level data on ITT. The industry cases assess the relative efficacy of local content rules (Chap. 11), joint ventures (Chap. 11), and offsets (Chap. 12) to achieve development objectives.

References

Baldwin W (1969) The case against infant industry protection. J Polit Econ 77(3):295–305

Bolton P (2006) Government procurement as a policy tool in South Africa. J Publ Procurement 6(3):193–217

Chang H-J (2008) Bad Samaritans: the myth of free trade and the secret history of capitalism. Bloomsbury Press, New York

Edler J, Georghui L (2007) Public procurement and innovation—resurrecting the demand side. Res Policy 36:949–963
Eliasson G (2010) Advanced public procurement as industrial policy. Springer, New York
Geroski P (1990) Procurement policy as a tool of industrial policy. Int Rev Appl Econ 4(2): 182–198
Grossman G (1989) Promoting new industrial activities: a survey of recent arguments and evidence, Papers 147. Woodrow Wilson School—Public and International Affairs, Princeton, pp. 87–125
Kattel R, Veiko L (2010) Public procurement as an industrial policy tool: an option for developing countries? J Publ Procurement 10(3)368–404
Li Y (2011) Public procurement as a demand-side innovation policy tool in china: a national level case study. Working Paper. http://druid8.sit.aau.dk/druid/acc_papers/oicntttfj00xg9i08f96sexh7y28.pdf. Accessed 10 June 2011
Lipsey RG, Lancaster K (1956) The general theory of second best. Rev Econ Studies 24(1):11–32
Luckey J (2009) The Buy American Act: requiring government procurements to come from domestic sources. http://www.seia.org/galleries/pdf/CRS_Report_-_The_Buy_American_Act_3.13.09.pdf. Accessed 25 June 2011
Pack H, Saggi K (2006) Is there a case for industrial policy? A critical survey. World Bank Res Obs 21(2):267–297
Reinert E (2008). How rich countries got rich and why poor countries stay poor. Public Affairs, New York.
Rodrik D (2008) Is there a new Washington consensus? http://www.project-syndicate.org/commentary/rodrik20/English. Accessed 12 July 2011
Spence M (2011) The next convergence. Farrar, Straus and Giroux, New York
Uryu K (2006) Government procurement as industrial policy: in support of Japan's defense aircraft, start-up and venture companies, and information technology sectors. USJP Occasional Paper 06–14. http://www.wcfia.harvard.edu/us-japan/research/pdf/06-14.urya.pdf. Accessed 10 June 2011
Watermeyer R (2000) The use of targeted procurement as an instrument of poverty alleviation and job creation in infrastructure projects. Publ Procurement Law Rev 2(5):226–250
Williamson J (1990) What Washington means by policy reform. In: Williamson J (ed) Latin American adjustment: how much has happened? Institute for International Economics, Washington, DC
World Bank (1993) The East Asian miracle. Oxford University Press, Oxford

Contents

Contributors

K. Ali Akkemik Kadir Has University, Istanbul, Turkey

Ulli Arnold University of Stuttgart, Stuttgart, Germany

Houssam-Eddine Bessam University of Stuttgart, Stuttgart, Germany

Fuquan Sun Institute of Comprehensive Development (CASTED), Beijing, People's Republic of China

Rainer Gadow University of Stuttgart, Stuttgart, Germany

Richard Haines Nelson Mandela Metropolitan University, Port Elizabeth, South Africa

Muammer Kantarci Turkish Coach Industry Incorporation (TÜVASAŞ), affiliated company of TCDD, Adapazari, Turkey

Ahmet Kesik Ministry of Finance, Ankara, Turkey

Mahmut Kiper Technology Development Foundation of Turkey (TTGV), Ankara, Turkey

Fuat Oktay Turkish Airlines Technic Inc, Istanbul, Turkey

Vehbi Özer Turkish Airlines Technic Inc, Istanbul, Turkey

Travis K. Taylor Christopher Newport University, Newport News, VA, USA

Murad Tiryakioğlu Afyon Kocatepe University, Afyonkarahisar, Turkey

Wanjun Deng Institute of Comprehensive Development (CASTED), Beijing, People's Republic of China

Ron Watermeyer Soderlund and Schutte Ltd, Johannesburg, South Africa

Yalçın Yılmazkaya Kale Group Companies Inc, Istanbul, Turkey

Murat A. Yülek THK University, Ankara, Turkey

Part I
Public Procurement Policies, Industrialization and Technology Transfer

Chapter 1
Public Expenditures on Machinery and Equipment in Developing Countries: A Potential Driver of Technological Development and Industrialization

Murat A. Yülek

Introduction

Growth performance in developing economies has recently been quite impressive. In parallel, fiscal health has improved. These factors are driving public and private spending capacity in these economies.

This chapter attempts at estimating and projecting (1) public procurement outlays devoted to machinery and equipment in these economies; and (2) importations arising from these outlays. Results of these estimations are likely to shed light on the negotiation power of developing countries in the international markets. Moreover, the results can shed light on their power to benefit from these outlays to support local technological development.

Well structured procurement policies capitalizing on the negotiation power can be instrumental in converting these economies from passive importers of technology products to manufacturers of such imported goods, bringing in value added to local economies and assisting in building local learning curves in developing countries.

Public Expenditure in Developing Countries

Developing countries are characterized by inadequate physical infrastructure. That, in turn, negatively affects the quantity and quality of their public services and hampers economic competitiveness. As a result, this low base of physical infrastructure in developing countries, coupled with generally high population growth and large territories necessitate large public expenditures.

M.A. Yülek (✉)
THK University, Ankara, Turkey
e-mail: myulek@pglobal.com.tr; mayulek@thk.edu.tr

M.A. Yülek and T.K. Taylor (eds.), *Designing Public Procurement Policy in Developing Countries*, DOI 10.1007/978-1-4614-1442-1_1,

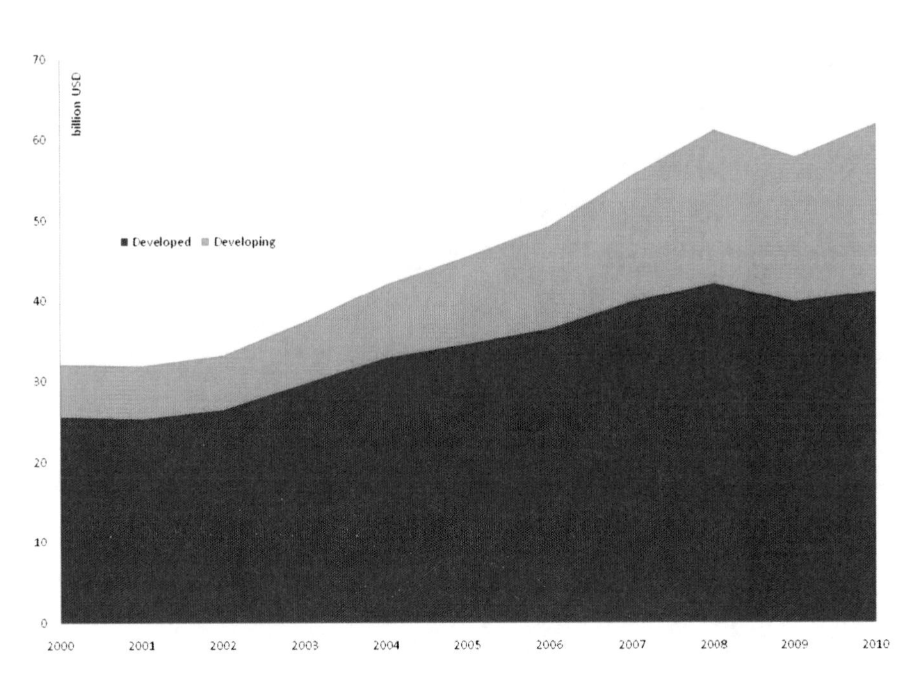

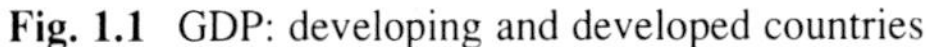
Fig. 1.1 GDP: developing and developed countries

In most cases, except for countries that are well-endowed with natural resources, the need for capital expenditures in developing countries is constrained by their limited capacity to generate public revenue. Recent trends have, however, suggested that these constraints are becoming relatively less significant.

The aggregated GDP of developing countries reached USD 20.8 trillion in 2010 from USD 6.5 trillion in 2000 (Fig. 1.1). That corresponds to an average nominal growth rate of 12.4% per annum, while the average nominal growth rate of GDP for developed countries was 4.8% during the same period. As a direct consequence, the share of developing countries in the world GDP increased from 20% in 2000 to 34% in 2010. Under currently prevailing trends, that growth differential between developed and developing countries will continue in the coming decades, further increasing the importance of the developing economies and their weight in the global economy.

In developing countries, growing economic activity translates into a larger base for public revenues and therefore a higher capacity for public expenditure. While the ratio of public expenditure outlays to total GDP has remained within a range of 4–5% of GDP in the last few years, growing GDP has created a larger base for public expenditures, as discussed above.

Strong fiscal balances represent another salient trend to coincide with the growing economic activity in developing countries. Due to this, the economies of developing countries were hit less severely than developed economies during the recent global financial crises. Furthermore, fiscal health has helped maintain sustainable growth in the public expenditures of developing countries.

Fiscal health also creates an environment in developing countries that is more conducive to directing fiscal policy toward developmental needs compared to earlier decades. Newly developed instruments such as Public-Private-Partnership (PPP) schemes and increased capital flows from developed economies, as well as capital flows from within the community of developing nations, have also been instrumental in helping developing countries to increase their capacity for procurement.

As a result of these factors, over the last 5 years, public expenditures in developing countries have significantly increased. In a selected sample of such countries (Tables 1.1 and 1.2), expenditures by central governments increased more than twofold, from USD 188 billion in 2000 to USD 438 billion in 2010. Note that these figures cover only the central government expenditures and thus do not include expenditures by local administrations.

This trend of rapidly increasing public expenditure in developing countries is likely to be sustained over both the medium and long term due to both push

Table 1.1 Central government capital expenditure (billion USD)

	2006	2007	2008	2009	2010
Afghanistan	0.05	0.03	0.04	0.04	0.05
Argentina	-do-	-do-	-do-	-do-	-do-
Azerbaijan	1.9	3.21	6.1	5.5	5.1
Bangladesh	0.0	2.96	2.9	4.0	4.6
Brazil	-do-	-do-	-do-	-do-	-do-
Bulgaria	1.5	2.49	3.2	3.3	2.6
Chile	2.6	3.45	3.9	5.3	5.4
China	59.7	66.39	144.6	164.5	224.1
Egypt	3.7	4.95	6.8	5.8	5.9
Georgia	0.7	0.88	0.9	0.8	0.9
Ghana	-do-	-do-	-do-	-do-	-do-
Hungary	4.9	4.98	4.0	3.9	4.6
India	13.6	17.27	27.7	18.6	20.0
Kazakhstan	4.0	6.39	8.3	6.8	8.6
Kenya	1.1	1.91	2.0	2.7	0.3
Korea	27.6	35.67	33.5	35.0	39.5
Malaysia	9.1	10.47	12.2	13.9	14.2
Mauritania	0.0	0.24	0.2	0.2	0.0
Morocco	0.0	-do-	-do-	-do-	-do-
Nigeria	6.3	0.50	9.1	6.4	11.2
Pakistan	-do-	-do-	5.1	5.7	5.9
Poland	11.3	15.31	28.6	24.6	32.5
Romania	5.0	8.19	9.4	7.1	6.2
Russia	-do-	-do-	-do-	-do-	-do-
Senegal	0.9	0.13	0.1	0.1	0.1
Singapore	4.2	4.95	7.9	9.1	12.2
South Africa	3.4	3.72	5.8	6.3	8.9
Taiwan	-do-	-do-	-do-	-do-	-do-
Thailand	7.9	9.64	10.6	11.6	...
Tunisia	2.1	2.30	2.6	2.9	2.9

(continued)

Table 1.1 (continued)

	2006	2007	2008	2009	2010
Turkey	10.1	9.74	13.9	12.9	13.1
Uganda	0.6	0.67	1.1	1.4	1.2
Uruguay	0.5	0.67	0.7	1.1	1.4
Vietnam	5.5	6.47	7.3	6.3	6.5
Memo					
Total	188.1	223.58	358.8	365.8	438.0
Average	6.7	8.28	12.8	13.1	16.2
Median	3.5	3.7	6.0	5.7	5.9

Source: IMF

Table 1.2 Central government capital expenditure (percent of GDP)

	2006	2007	2008	2009	2010
Afghanistan	0.6	0.3	0.3	0.3	0.3
Argentina	-do-	-do-	-do-	-do-	-do-
Azerbaijan	9.0	9.7	13.2	12.7	9.8
Bangladesh	-do-	4.0	3.4	4.2	4.4
Brazil	-do-	-do-	-do-	-do-	-do-
Bulgaria	4.7	6.3	6.4	7.1	5.9
Chile	1.8	2.1	2.3	3.3	2.7
China	2.2	1.9	3.2	3.3	3.9
Egypt	3.4	3.8	4.2	3.1	2.7
Georgia	9.0	8.6	7.9	7.7	8.0
Ghana	-do-	-do-	-do-	-do-	-do-
Hungary	4.3	3.6	2.6	3.0	3.5
India	1.5	1.5	2.2	1.5	1.4
Kazakhstan	4.9	6.2	6.1	6.3	6.6
Kenya	4.6	6.6	7.6	8.9	1.0
Korea	2.9	3.4	3.6	4.2	4.0
Malaysia	5.8	5.6	5.5	7.2	6.5
Mauritania	-do-	8.6	7.0	6.8	1.1
Morocco	-do-	-do-	-do-	-do-	-do-
Nigeria	4.3	5.9	4.4	3.8	5.4
Pakistan	-do-	-do-	3.1	3.5	3.4
Poland	3.3	3.6	5.4	5.7	7.4
Romania	4.1	4.8	4.6	4.4	3.9
Russia		-do-	-do-	-do-	-do-
Senegal	9.8	1.1	1.0	1.0	1.1
Singapore	2.9	2.8	4.1	5.0	5.6
South Africa	1.3	1.3	2.1	2.2	2.5
Taiwan	-do-	-do-	-do-	-do-	-do-
Thailand	3.8	3.9	3.9	-do-	-do-
Tunisia	6.1	5.9	5.8	6.6	6.6
Turkey	1.9	1.5	1.9	2.1	1.8
Uganda	5.7	5.6	7.7	8.6	7.2
Uruguay	2.6	2.8	2.1	3.6	3.4
Vietnam	9.1	9.1	8.1	6.8	6.4
Memo					
Average	4.4	4.5	4.6	4.9	4.3
Median	4.1	3.9	4.2	4.2	3.9

Source: IMF

(economic growth, better public finances and better external financing resources) and pull (the need for rapid development of physical infrastructure) factors.

Public Expenditure, Share of Machinery and Equipment in Public Expenditure and Imports of Machinery and Equipment: The Case of Turkey

Public expenditure consists of current as well as capital expenditures. The former, which concerns day-to-day needs of the economy, generally takes the lion's share of total expenditures. In cases where the public balance sheet has deteriorated, current expenditures take an even larger share. For example, debt overhang in developing countries has caused prohibitive debt service costs, which have squeezed resources that could be allocated to capital expenditures. Nevertheless, as discussed in the previous section, capital expenditures have increased considerably recently.

Table 1.3 presents a summary of statistics on Turkey's central government capital expenditures, which increased considerably from USD 54 billion in 2000 to USD 150 billion in 2010 despite a significant negative spike during the 2001 macroeconomic and financial crises. This translates into an investment rate of approximately 20% of GDP, excluding the years of slowdown emanating from internal (2001–2002) or external (2009) factors. These rates place Turkey among countries with lower investment levels compared to Asian economies, but at median levels among developing economies overall.

The public sector has generally been responsible for less than 20% of the country's capital expenditures. This has resulted in an increase of investment from USD 8 to 25 billion per annum between 2005 and 2010, equivalent to an average of 3.8% of GDP.

Machinery and equipment constituted, on average, 55% of the country's total capital expenditure; the remaining part represents building and construction. In the central government, machinery and equipment represented 21% of total capital expenditure on average, while in the private sector it was 62%. Thus, four-fifths of public expenditures consisted of construction work. As a share of GDP, central government expenditure on machinery and equipment has stabilized over time from 1.3% in 2000 to 0.6% after 2005.

An open economy, Turkey imports a large quantity of goods, to the order of 25% of its GDP. Of this, typically 35% consists of various machinery and equipment,[1] equivalent to 9% of GDP.

[1] This figure includes the following items: machinery and transport equipment; prefabricated buildings; sanitary, plumbing, heating and lighting fixtures and fittings, professional, scientific, and controlling instruments; and apparatus, photographic apparatus, equipment and supplies; and optical goods, watches, and clocks.

Table 1.3 Turkey: public expenditure and capital expenditure

	2000	2001	2002	2003	2004	2005	2006
Total capital expenditure (billion USD)	54.3	31.1	38.8	51.7	79.8	101.5	117.8
Central government (billion USD)	12.4	8.3	10.1	10.3	11.5	15.9	17.9
Private (billion USD)	41.9	22.8	28.7	41.5	68.3	85.6	99.9
Percent of total							
Central government	22.8	26.5	26.0	19.9	14.4	15.7	15.2
Private	77.2	73.5	74.0	80.1	85.6	84.3	84.8
Total capital expenditure (percent of GDP)	20.4	15.9	16.7	17.0	20.3	21.0	22.3
Central government	4.6	4.2	4.3	3.4	2.9	3.3	3.4
Private	15.7	11.7	12.4	13.6	17.4	17.7	18.9
Total central government capital expenditure on machinery and equipment (billion USD)	30.6	15.3	19.9	28.2	47.2	60.4	67.4
Central government	3.5	2.1	2.9	2.5	2.7	3.8	3.2
Private	27.1	13.2	17.0	25.7	44.5	56.7	64.2
Percent of total							
Central government	11.4	13.8	14.5	8.8	5.8	6.2	4.7
Private	88.6	86.2	85.5	91.2	94.2	93.8	95.3
Percent of GDP							
Central government	1.3	1.1	1.2	0.8	0.7	0.8	0.6
Private	10.2	6.8	7.3	8.5	11.3	11.7	12.2
Percent of total central government capital expenditure	56.3	49.3	51.4	54.5	59.2	59.5	57.2
Central government (percent of total central government capital expenditure)	28.2	25.6	28.6	24.1	23.7	23.6	17.7
Private (percent of total private capital expenditure)	64.5	57.9	59.3	62.1	65.1	66.2	64.3
Machinery and equipment imports (billion USD)	21.9	13.7	16.8	23.0	35.8	40.7	46.0
Non-automobile	16.5	11.9	14.5	17.7	25.7	30.3	34.8
Automobile	5.4	1.8	2.3	5.3	10.1	10.4	11.1
Central government (billion USD)/1	2.5	1.9	2.4	2.0	2.1	2.5	2.2
Private (billion USD)/1	19.4	11.8	14.4	21.0	33.8	38.2	43.8
Central government (percent of GDP) /1	0.9	1.0	1.0	0.7	0.5	0.5	0.4
Private (percent of GDP) /1	7.3	6.1	6.2	6.9	8.6	7.9	8.3
Memo							
Machinery and equipment imports (percent of total imports)	40.2	33.1	32.6	33.2	36.7	34.8	32.9
Non-automobile	30.2	28.7	28.1	25.5	26.4	25.9	24.9
Automobile	10.0	4.4	4.5	7.7	10.4	8.9	8.0
Total imports (billion USD)	54.5	41.4	51.6	69.3	97.5	116.8	139.6
GDP (billion USD)	266.3	195.1	232.1	304.1	392.2	482.8	528.6

Source: Turkish Statistical Institute (National Accounts Statistics); calculations by the author
1. Calculated based on the share of central government machinery and equipment expenditure in total

How much of these machinery and equipment importations are accounted for by the government procurement? Government statistics, in many other developing countries (as well as developed countries), do not report this figure. However, based on the central government's share of total capital expenditure on machinery and equipment, we can infer a ratio of 0.5–1% for Turkey (Table 1.3). This is calculated by multiplying imports of machinery and equipment with the ratio of central government's machinery and equipment expenditures to total. Note that the procurements of such goods by local administrations are not included in this.

Imported Public Expenditures in Developing Countries on Procurement of Machinery and Equipment: Estimates and Projections

In many developing countries, public outlays for procurement of machinery and equipment are generally either not reported, or access to that information is constrained for various reasons. In fact, this is true for most economies: import statistics on publicly procured machinery and equipment are not reported.[2] It is therefore necessary to estimate these outlays using certain constructs as guidelines. In this section, a simple guideline is employed based on the statistics from previous sections.

The guideline developed here uses GDP as an appropriate indicator of the size of an economy and as a basis for estimating public capital expenditures in developing countries. International Monetary Fund (IMF) data is used for this purpose; IMF's World Economic Outlook (WEO) database is considered a reliable database for various macroeconomic statistics, including GDP. Moreover, WEO presents the projections of IMF staff for individual country GDP levels in USD terms for the 5 subsequent years. WEO data is updated regularly twice a year. The data and projections used in this section are taken from the statistics published in April 2011.

Developing countries listed in Table 1.4 include all countries that IMF classifies as "emerging and developing countries" in WEO. Some of the countries that IMF classifies as "advanced countries" could, in point of fact, be considered among the developing countries[3]; however, for purposes of consistency we have not done so (Fig. 1.2).

We have used lower and upper brackets for the ratio of public expenditures on machinery and equipment to GDP. The Turkish case, presented in the previous

[2] Publicly procured machinery and equipment may sometimes be imported directly. In other cases, publicly procured domestic machinery and equipment may be aggregated in other figures for imported goods. To the author's knowledge no country, developed or developing, reports the direct or indirect import of publicly procured machinery and equipment.

[3] Such as Cyprus, Estonia, Greece, Portugal, Slovak Republic, and Slovenia.

Table 1.4 List of developing economies included in the estimates

Afghanistan	Georgia	Pakistan
Albania	Ghana	Panama
Algeria	Grenada	Papua New Guinea
Angola	Guatemala	Paraguay
Antigua and Barbuda	Guinea	Peru
Argentina	Guinea-Bissau	Philippines
Armenia	Guyana	Poland
Azerbaijan	Haiti	Qatar
The Bahamas	Honduras	Romania
Bahrain	Hungary	Russia
Bangladesh	India	Rwanda
Barbados	Indonesia	Samoa
Belarus	Islamic Republic of Iran	São Tomé and Príncipe
Belize	Iraq	Saudi Arabia
Benin	Jamaica	Senegal
Bhutan	Jordan	Serbia
Bolivia	Kazakhstan	Seychelles
Bosnia and Herzegovina	Kenya	Sierra Leone
Botswana	Kiribati	Solomon Islands
Brazil	Kosovo	South Africa
Brunei Darussalam	Kuwait	Sri Lanka
Bulgaria	Kyrgyz Republic	St. Kitts and Nevis
Burkina Faso	Lao People's Democratic Republic	St. Lucia
Burundi	Latvia	St. Vincent and the Grenadines
Cambodia	Lebanon	Sudan
Cameroon	Lesotho	Suriname
Cape Verde	Liberia	Swaziland
Central African Republic	Libya	Syrian Arab Republic
Chad	Lithuania	Tajikistan
Chile	Former Yugoslav Republic of Macedonia	Tanzania
China	Madagascar	Thailand
Colombia	Malawi	Timor-Leste
Comoros	Malaysia	Togo
Democratic Republic of Congo	Maldives	Tonga
Republic of Congo	Mali	Trinidad and Tobago
Costa Rica	Mauritania	Tunisia
Côte d'Ivoire	Mauritius	Turkey
Croatia	Mexico	Turkmenistan
Djibouti	Moldova	Tuvalu
Dominica	Mongolia	Uganda
Dominican Republic	Montenegro	Ukraine
Ecuador	Morocco	United Arab Emirates
Egypt	Mozambique	Uruguay
El Salvador	Myanmar	Uzbekistan
Equatorial Guinea	Namibia	Vanuatu
Eritrea	Nepal	Venezuela
Ethiopia	Nicaragua	Vietnam
Fiji	Niger	Republic of Yemen
Gabon	Nigeria	Zambia
The Gambia	Oman	Zimbabwe

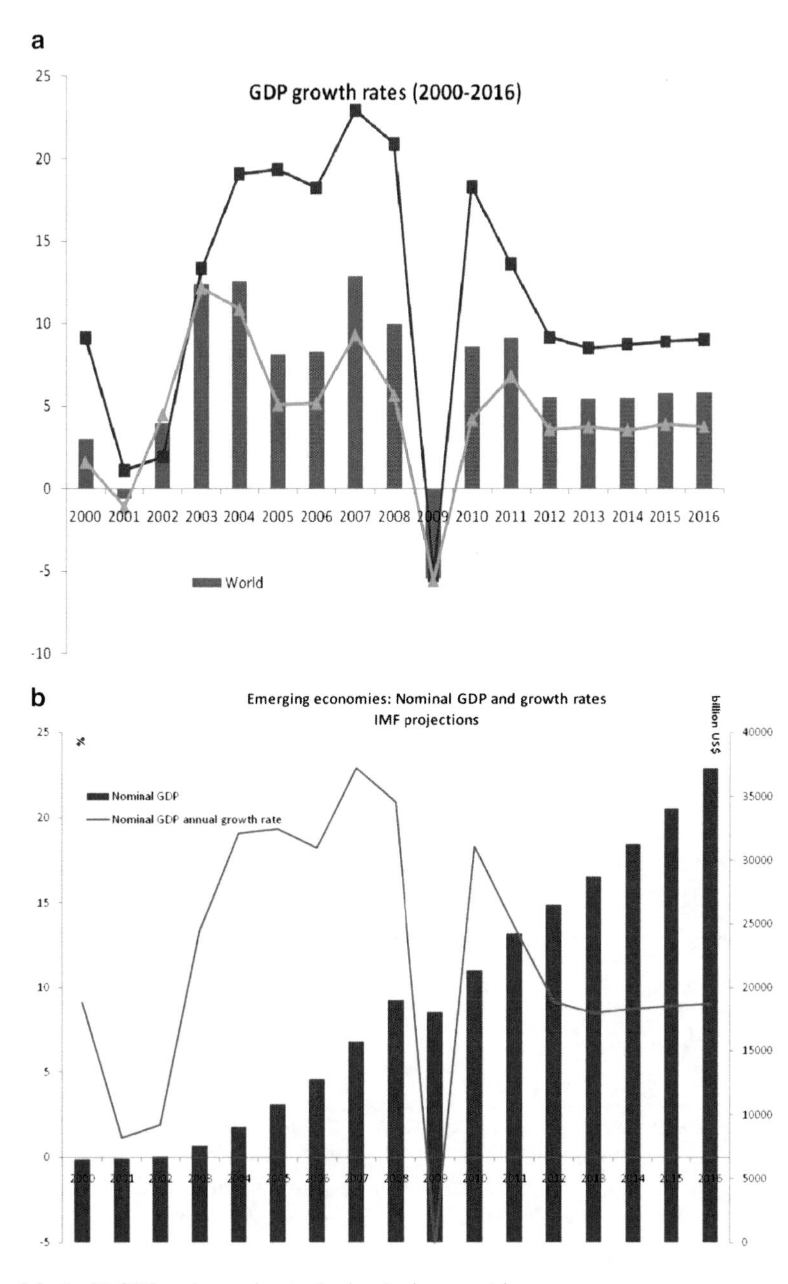

Fig. 1.2 (**a**, **b**) GDP and growth rates in developing countries

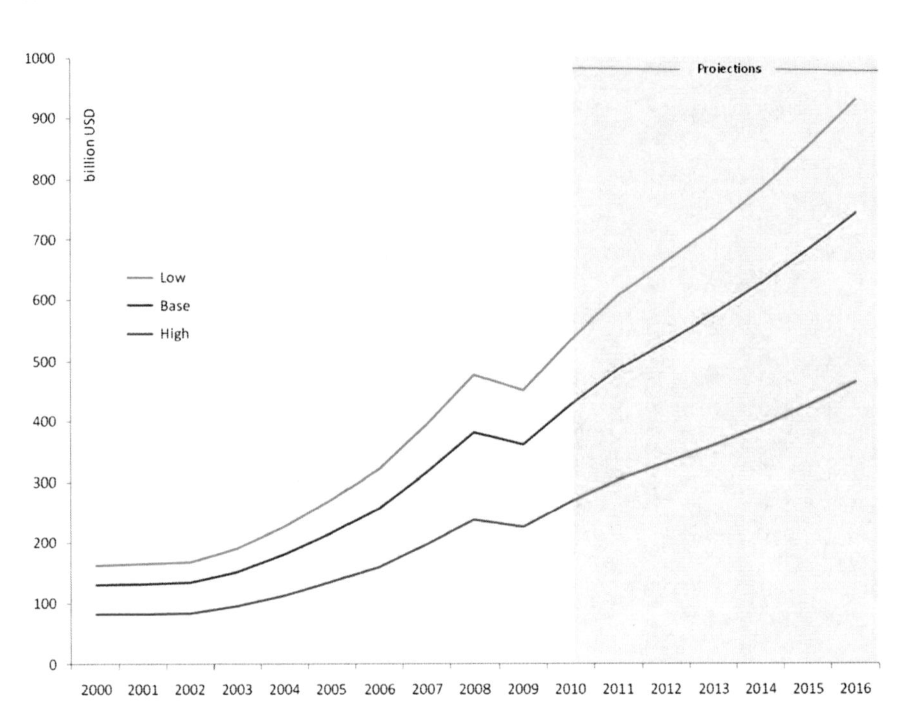

Fig. 1.3 Public procurement of machinery and equipment in emerging economies: estimates and projections

Table 1.5 Projected estimates for imports of publicly procured machinery and equipment in developing economies (billion USD)

	2005	2010	2011	2012	2013	2014	2015	2016
High	135.6	267.2	303.6	331.5	359.7	391.2	426.1	464.6
Base	81.4	160.3	182.2	198.9	215.8	234.7	255.6	278.8
Low	54.2	106.9	121.4	132.6	143.9	156.5	170.4	185.8

Source: author' calculations

section, results in a ratio of 0.5–1% of GDP. In this section, we use a slightly wider range of 0.5 ("Low") percent to 1.25% ("High"), to allow for developing countries with more relaxed budget constraints. We have also included a base case at 1% of GDP. Figure 1.3 presents the estimated ranges projected to 2016 for total central government expenditure on machinery and equipment in the developing countries listed in Table 1.4. In Table 1.5, the range figures are presented.

Our estimate shows that by 2016, governments of developing countries will be spending between USD 186 to USD 465 billion on purchases of machinery and equipment. These figures do not include procurement made through PPP projects as well as central government expenditure of local administrations.

Conclusions

Various factors suggest that public procurement of machinery and equipment in developing countries will continue to grow significantly in the future. The process is supported by both pull and push factors.

On the push side, developing countries have registered high growth rates in the last decades. This growth performance is likely to be sustained in the medium to long term and will provide a basis for increased generation of public revenues in developing countries, which can be used to finance public procurement in general and the purchase of machinery and equipment in particular. Second, fiscal health in the developing countries will present an environment conducive to increased public spending.

On the pull side, the increasing size of the economies in developing countries, their large land areas, and substantial population bases require sizeable public expenditures. Existing conditions in these countries, on the other hand, point to a weak base of physical infrastructure that will necessitate large amounts of green-field infrastructure investments in addition to the usual upgrading, reconstruction, and rehabilitation investments.

Pull and push factors, combined, have led to public spending on significant amounts of machinery and equipment. In turn, these countries have become significant importers of machinery and equipment. Our estimates indicate that these imports will attain a range of USD 186 billion to USD 465 billion per annum by 2016, granting developing countries significant negotiating power vis-à-vis exporters of machinery and equipment. It has to be added that, if coordination can be achieved with private importers, negotiating power would increase immensely.

Chapter 2
Countertrade Offsets in International Procurement: Theory and Evidence*

Travis K. Taylor

Introduction

International government procurement is rife with deviations from the perfectly competitive market model and arm's-length exchange. In the defense, aerospace, capital equipment, automotive, and telecommunications industries, for example, we encounter imperfectly competitive markets. The complexities of high transaction costs, incomplete and asymmetric information, and bounded rationality mark this exchange setting. In this environment, economic theory predicts that markets typically underproduce relative to the socially optimal level, and sellers frequently earn supernormal profits, rents, and quasi-rents.

Production is characterized by high start-up costs, fixed costs, and research and development (R&D) investments. The increasing returns associated with large-scale production runs make it critical to augment the domestic market with export sales. The mode of competition can vary among sellers in these markets by industry. Standard oligopoly models predict positive profits, though this need not always be the case. One can envision scenarios where two firms are the principal sellers of a product, and Bertrand competition ensues. In such circumstances, intense competition could drive price down to the perfectly competitive level with both firms earning economic profits equal to zero.

*This chapter was originally published in Thai, K. (ed.), *New Horizons in Public Procurement* (2011), PrAcademics Press, Florida USA. The editors are grateful to PrAcademics Press for their cooperation and reprint authorization. The author gratefully acknowledges funding support from a Christopher Newport University research grant. Mr. Andrew Winters provided excellent research assistance. The author also thanks three referees who provided valuable comments and suggestions. Any remaining errors are, of course, the sole responsibility of the author.

T.K. Taylor (✉)
Christopher Newport University, Newport News, VA, USA
e-mail: ttaylor@cnu.edu

M.A. Yülek and T.K. Taylor (eds.), *Designing Public Procurement Policy in Developing Countries*, DOI 10.1007/978-1-4614-1442-1_2,

For example, consider the world market for surface-to-air missile-defense systems. Raytheon (US) produces the Patriot missile, and the Russian firm Almaz makes the S-300. In the absence of collusion, Bertrand competition may ensue. To avoid an all out price war, the firms may attempt to influence the terms of trade beyond the standard price margin. Bribery, favoritism, and other dubious measures are illegal and thus discouraged. Barter, countertrade, and offset contracts are legal to varying degrees. These procurement policies can be understood as marketing strategies to differentiate the oligopolists' goods. As we analyze the demand side of these transactions, we find that countertrade offsets are attractive for a variety of perceived and real benefits. In instances when the arrangement matches sellers with previously unknown buyers and input providers, production and transaction costs can be reduced (Taylor 2005). Others (Brauer 2004; Taylor 2004; Martin 1996) note that procurement-induced countertrade can foster technology transfer, conservation of foreign exchange, market penetration, foreign investment, training and services, and more.[1]

In this chapter, we examine the rationale for nonstandard contracting in international procurement. Long viewed as trade diverting and inefficient, we assess the benefits and costs of countertrade offsets and explain their very existence in section one. Section two provides a legal context including the governing statutes of the World Trade Organization (WTO), European Union (EU), and the US government. In the third section, we focus on the transaction costs of monitoring the contracts, *ex post*. This section notes that measurement costs range from negligible to significant and monitoring regimes need to adapt accordingly. We evaluate the settings in which input metrics, output metrics, and some combination of the two are advisable. The final section of the chapter reports several empirical findings drawn from a new countertrade database.

Theoretical Rationale for Countertrade Offsets in Procurement

An offset agreement is a contract between a purchasing government and a foreign supplier. As a condition for the sale of goods or services (the "base good"), the foreign firm is encouraged or even required to provide additional economic benefits – beyond the base transaction – to the purchasing government's economy. These benefits can take the form of countertrade, industrial compensation packages, investment, technology transfer, subcontracting, and so forth. In essence, the offset inserts a degree of reciprocity in the transaction. The perceived benefits of procurement countertrade, while politically attractive, are difficult to verify empirically. First, the benefits of the

[1] According to the *U.S. Presidential Commission on Offsets* (2001, p. 32), seven of eight large American aerospace firms reported that they would lose 50–90% of their export sales if offsets were not included in the deal.

offset must be distinguished from economic activity that may have occurred naturally in the absence of the intervention. This requirement is known as additionality.[2] Second, the benefits accrue over a long period, rather than a lump sum monetary transfer. According to the U.S. Commerce Department's annual *Offsets in Defense Trade* (2007), the average time to fulfill the offset obligation is approximately 7.5 years. This duration of time makes net benefit calculations more difficult and unreliable – though certainly still useful. Assumptions about the depreciation of core capabilities, technology, and other time-sensitive assets require codification to maintain best practices.

Third, the purchasing government needs to compute the economic cost of the offset arrangement, which is no easy task. Economic costs include the opportunity cost of resources that are redirected toward the offset, as well as the additional variable and fixed costs to fulfill the offset. The extra variable and fixed costs are usually passed on to the buyer – at least partially – in the form of a higher price of the base good.[3]

In short, there is much skepticism in the literature whether offset arrangements, on balance, show positive net benefits for a country.[4] In one of the few economic audits ever performed and released to the public, PriceWaterhouseCoopers (PWC) was hired to perform an independent study of several representative offset agreements signed by the government of the Netherlands (Countertrade and Offset 2003). Though the purchase price increased nominally on average, the net benefits were positive. Such studies are important and ought to become common practice.

Another set of questions emerges when governments elect to forego price discounts and instead opt for countertrade arrangements in procurement. In government procurement markets, there are typically a relatively small number of buyers, each possessing significant purchasing power. In industries such as information technology, aerospace, telecommunications, and defense, we note a highly concentrated market structure. The combination of few buyers and few sellers creates a bilateral oligopsony market structure that induces bargaining and a wide range of transaction prices, particularly when pricing data are proprietary and seldom released.[5]

Most interesting – and troubling to many at first blush – is the request for in-kind transfers in lieu of price discounts. In Arrow and Debreu markets competitive forces create an equilibrium price that equals the marginal cost and long run

[2] To fulfill offset obligations, most sellers are asked to demonstrate additionality and conditionality. Additionality refers to new economic activity that was transferred from the seller to the purchasing government's economy, above any activity that may have occurred in free marketplace. A related term is conditionality, which connotes the reciprocal exchange among multiple firms.

[3] This practice is known as "price padding." Taylor (2005) and PriceWaterhouseCoopers (Countertrade and Offset 2003) estimate that price padding ranges from 3 to 5% of the original purchase price.

[4] See Brauer (2004), for a discussion of offset audits.

[5] If the firm can withhold pricing data in a highly concentrated market, it may be able to price discriminate and increase profits.

average cost of production.[6] Competition ensures the lowest possible price, while bargaining and its associated transaction costs are assumed to be negligible or zero. As we move toward imperfectly competitive markets, economic models predict economic rents for sellers and discounted prices for large buyers. Depending on the extent of market power, the purchasing government may negotiate discounts that extract rents from the seller. Why, then, do purchasing governments opt for in-kind transfer instead of cash? Several compelling theories have been offered in the literature. Hall and Markowski (1994, 2004) and Taylor (2004) examine two broad categories: (1) neoclassical cost minimization, and (2) the capabilities theory of the firm.

Hall and Markowski (1994) note that procurement of large-scale, high technology products such as weapons systems, aerospace, and information technology typically involves thousands of complementary products. For example, when Northrup Grumman sells its aircraft carriers, it also offers related items such as "acquisition management services," "acquisition support," "active tracker laser," and the "LN-120G Stellar-Inertial Navigation System" that are typically purchased as well.[7] In short, international government procurement is seldom the stand-alone, turnkey variety. To operate properly, systems require significant training, maintenance, and service after the sale.

As a result, bundling complementary products together may be Pareto-efficient for all parties. When the seller bundles offset work with the base good (the aircraft carrier in the example above), it may achieve economies of scale and scope.[8] However, offset demands that are outside of the seller's core capabilities may extend the boundary of the firm and subject it to diseconomies of scope. Poorly conceived countertrade agreements and indirect offsets are most likely to suffer from these diseconomies and increased transaction costs.

The potential for consumption economies also exists. For example, if Spain were to purchase a nuclear submarine from Northrop Grumman, tremendous cost savings can accrue if Spain uses the same firm to build the accompanying dock, provide maintenance, and *ex post* service. Rather than working with multiple firms and signing numerous contracts, the buyer can exploit the complementarities in consumption and reduce transaction costs by signing a multiyear offset agreement. Probably the most important aspect of this arrangement, however, is that the buyer extracts rents while also augmenting its capabilities. Evidently, the procurement via countertrade is perceived to be more successful in this regard than arm's-length exchange in markets.

[6] Arrow and Debreu (1954) markets are characterized by many buyers and many sellers, no barriers to entry, product homogeneity, marginal cost pricing, and complete information.

[7] See http://www.atoz.northropgrumman.com/Automated/AtoZ/L.html.

[8] Economies of scale occur when long run average cost per unit declines as more is produced. Mathematically, economies of scale is given by $\partial LRAC/\partial Q < 0$.Economies of scope are achieved when the cost of producing two different products together is less than the cost of producing them separately ($C(y1,y2) < C(y1) + C(y2)$), where C is total cost, y1 is the output of product 1 and y2 is the output of product 2.

Purchasing governments may elect to leave the price margin for several reasons grounded in the economics of organization literature. First, offsets can be designed to safeguard the base good when the seller is required to post an economic hostage (Williamson 1983; Hennart 1989). For example, one technique is to sign a coproduction or subcontracting agreement whereby the seller is responsible for collaborating with a domestic firm in the purchasing government's economy. The requirement is to produce inputs that are then sold back to the seller and used in the base good's production function. Inasmuch as the offset creates incentive for the seller to teach local firms best practices and cost-minimizing techniques, incentives are better aligned. This countertrade strategy is particularly valuable in cases where the base good embodies tacit knowledge and high technological intensity in production (Penrose 1959).

Second, the offset may serve as the carrot needed to induce a multinational firm to sell, market, teach, advise, or invest in a third country. To be clear, the seller is unlikely to give away its core capabilities and in many instances the American Bureau of Export Administration at the Commerce Department makes it illegal to transfer certain products to select countries and end-users. Nevertheless, there is anecdotal evidence that purchasing governments are able to assist domestic firms in obtaining goods and services that would otherwise not be forthcoming in free markets.[9] Ultimately, we return to the basic notion of the "make-or-buy" decision. Is it more efficient for a government to purchase a product off-the-shelf in free markets, or use its leverage to induce domestic production via countertrade and offsets? This is not an easy question to answer, especially because of the intertemporal dynamics involved. Competencies and comparative advantages are not static; they are dynamic and can be learned. The challenge is to determine which method is most efficient to obtain the desired results.

Legal Framework

The Agreement on Government Procurement (GPA) of the WTO Uruguay round (1994) addresses the issues of international procurement, countertrade, and offsets. Article XVI sets forth the official WTO position on offsets:

- Entities shall not, in the qualification and selection of suppliers, products, or services, or in the evaluation of tenders and award of contracts, impose, seek, or consider offsets.
- Nevertheless, having regard to general policy considerations, including those relating to development, a developing country may at the time of accession negotiate conditions for the use of offsets, such as requirements for the incorporation of domestic content. Such requirements shall be used only for

[9] Interviews with U.S. Commerce Department and Malaysian Ministry of Finance, 1998.

qualification to participate in the procurement process and not as criteria for awarding contracts. Conditions shall be objective, clearly defined, and nondiscriminatory.

Article XVI is a plurilateral agreement, thus members are encouraged but not required to be a signatory to it. In 2010, 40 of the 153 WTO members are signatories to this agreement (World Trade Organization 2010). Broad consensus exists for free market competition and transparency of procurement policies.[10] Article XVI reflects this consensus by prohibiting countertrade requirements like offsets in civil procurement.

Several important exceptions to competition and transparency exist, however. First, developing countries are encouraged but not required to abide by this agreement.[11]

Article V: special and differential treatment for developing countries

Objectives

1. Parties shall, in the implementation and administration of this Agreement, through the provisions set out in this Article, duly take into account the development, financial and trade needs of developing countries, in particular least-developed countries, in their need to
   (a) Safeguard their balance-of-payments position and ensure a level of reserves adequate for the implementation of programmes of economic development
   (b) Promote the establishment or development of domestic industries including the development of small-scale and cottage industries in rural or backward areas; and economic development of other sectors of the economy
   (c) Support industrial units so long as they are wholly or substantially dependent on government procurement
   (d) Encourage their economic development through regional or global arrangements among developing countries presented to the Ministerial Conference of the World Trade Organization (hereafter referred to as the "WTO") and not disapproved by it

The rationale for exempting developing countries is akin to that of the infant industry protection theory. According to this theory, developing countries are home to a preponderance of new and inefficient firms. These firms have relatively high average costs of production relative to established multinational enterprises (MNEs) from industrialized countries. If these "infant" firms are forced to compete on the

[10] Linarelli (2003) details the WTO transparency agenda, and the endemic political and economic challenges that can stunt progress. Trionfetti (2003) reviews the theoretical literature and concludes that home-biased procurement can be trade-diverting (and output reducing) under monopolistic competition. Although the author does not address offset agreements, some types of countertrade contracts in procurement appear to fall in this category as well. On the other hand, some offset contracts can be shown (Taylor 2005) to lower unit costs, increase trade, and improve welfare. The wide variation of countertrade arrangements, then, limits the applicability of Trionfetti's (2003) findings to select settings.

[11] The exception in Article V is based on political and economic considerations. To achieve sufficient "buy-in" from a heterogeneous population of member countries, the exemptions in Article V proved necessary in the domestic political discourse. Moreover, economic development theory is far from conclusive in regard to the growth strategy for developing countries.

same platform as the MNEs in the GPA, they will invariably lose most of the time. Suppose, alternatively, that the government applies a temporary tariff on foreign imports of the industry in question. Proponents of this theory (Chang 2008; Reinert 2007) argue that when new firms have sufficient time to learn, innovate, and process tacit knowledge associated with the production process, average costs will fall and protection is removed as the "grown up" firm emerges. Although the economics literature is generally skeptical toward any purported net benefits of this policy, the theory (and policy) is attractive due to its perceived simplicity, logic, and political appeal (Baldwin 1969; Succar 1987).

A second set of exemptions from Article XVI concerns national security and public health. Defense purchases ranging from small arms and radars, to tanks and multibillion dollar weapons systems may include countertrade and offset requirements. Notably, this exemption applies to all countries.[12] As the reader may gather, an exemption for national security is so wide in scope that procurement officers can employ offset arrangements at their discretion. A similar exemption may be invoked when governments purchase goods and services for public health. To date, the vast majority of exemptions to Article XVI have come from defense procurement.

The European and American procurement policies complement the WTO's GPA. The European policy on defense procurement is codified in Article 296 of the EU's Economic Community (EC) treaty. Article 296 outlines a national security exemption that is consistent with Article XVI of the GPA. European governments have historically used countertrade and offsets to promote select industries and technologies in the defense industrial base (DIB). Recent discussions between Europe's G-6 and the US government signal a possible procurement policy shift toward less reliance on offsets. A 2009 white paper ("European Code of Conduct") from the European Defence Agency aims to develop a multilateral consensus on ways in which procurement offices can design offsets that support sustainable economic growth and limit adverse effects. Although the U.S. Inter Agency Working Group (IAWG) has clearly been the catalyst for these discussions, the Europeans have been willing to broach the topic far more than in the past. For instance, a new regulation in the EU will significantly limit the scope of activities that may be used to fulfill offset obligations. The European Defence Procurement Directive "targets indirect non-military offsets, with the aim of eliminating them" (U.S. Department of Commerce 2009, p. 52).

The US government rejects any claims that countertrade and offsets in procurement can yield positive net benefits. The official US policy is that countertrade and agreements are trade-diverting and inefficient. The government neither requires nor accepts offsets from foreign MNEs in RFPs and bid evaluations.[13] Furthermore, since

[12] Offsets may be included in the RFP, but they cannot be the determining factor in bid evaluation.

[13] An argument can be made that when the US government invokes the "Buy America" clause in many defense procurement transactions, the level of protection and trade-diversion far exceeds that of an offset or a related countertrade instrument.

Table 2.1 Summary of US defense offset arrangements, 1996–2008 (in $ millions)

Year	US companies	Agreements	Base export value[a] ($)	Offset value ($)	Offset ratio[b]%	Countries
1996	16	53	3,119.7	2,431.6	77.9	19
1997	15	60	5,925.5	3,825.5	64.6	20
1998	12	41	3,029.2	1,768.2	58.4	17
1999	10	45	5,656.6	3,456.9	61.1	11
2000	10	43	6,576.2	5,704.8	86.7	16
2001	11	34	7,017.3	5,460.9	77.8	13
2002	12	41	7,406.2	6,094.8	82.3	17
2003	11	32	7,293.1	9,110.4	124.9	13
2004	14	40	4,927.5	4,329.7	87.9	18
2005	8	25	2,259.8	1,464.1	64.7	18
2006	12	44	4,832.4	3,425.3	70.8	20
2007	10	43	6,735.7	5,437.5	80.7	18
2008	14	52	6,096.1	3,480.6	57.1	17

[a]*Base export value* is the summation of US sales to countries that require offsets
[b]*Offset ratio* is the compensation ratio defined as (offset value/base export value)
Source: U.S. Department of Commerce (2009, p. 5)

the Duncan Memorandum of 1978 the US government has (wisely) refused to guarantee any offset obligations that American firms owe to foreign countries. The government is, however, keenly aware that American firms dominate the very industries that are rife with countertrade and offset requirements. During the period of 1993–2008, 48 American firms signed 677 offset agreements worth $68.93 billion (U.S. Department of Commerce 2009, p. 4). In 2008 alone, 14 US companies entered into 52 offset agreements in support of $6.09 billion in export sales. Table 2.1 summarizes offset arrangements signed by US defense firms between 1996 and 2008.

The nominal value of offset obligations is a concern for the US government and the MNEs involved. As the nominal value rises relative to the value of the base sale, the long run profitability of MNE declines. During the last 15 years, the compensation ratio for American firms that signed offset agreements is 70.96% (U.S. Department of Commerce 2009, p. 5).[14] Thus, on average, every $1 in export sales in these markets is supported by about 71 cents in economic benefits returned to the purchasing country. These data are a bit misleading, however, because the offset fulfillment is usually spread over a 5–10-year period. In addition, Taylor (2005) shows that in imperfectly competitive markets with incomplete information, the offset can actually lower the MNEs' average cost of production under certain circumstances.

The government is also worried about the potential damage to second- and third-tier manufacturing firms due to countertrade and offset requirements. When the MNE ("prime contractor") fulfills its offset obligation by substituting a foreign input supplier

[14] The compensation ratio is defined as the dollar value of offset obligations divided by the dollar value of export sales supported by the agreement.

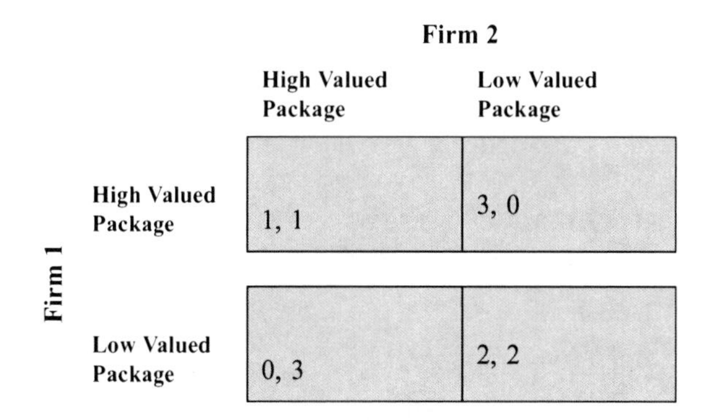

Fig. 2.1 Simple prisoner's dilemma game of offset provisions

for a domestic firm previously integrated in the supply chain, lower and mid-tier deindustrialization could occur. Empirical evaluation of this potential substitution effect is sparse and inconclusive.[15] The general equilibrium effects are ambiguous.

To address these and other concerns about offsets in government procurement, the IAWG was formed in 2004. The IAWG, which is comprised of officials from Commerce, Defense, and State is charged with engaging other offset-providing country governments to limit the growth of offsets. The IAWG also seeks to lessen the importance of offsets in bid evaluation, and promote increased flexibility toward offset fulfillment among purchasing governments. Meetings have been held with European counterparts, and policy coordination remains a remote possibility in the future.

Policy coordination in these markets will likely prove difficult. A stylized example can illustrate the resistance to policy coordination in a sealed bid, simultaneous game setting. Firms and governments operate in a bilateral oligopsony market structure. A handful of large firms – primarily from the US and Europe – compete for the business of a relatively small number of purchasing governments. In this simplified case, assume that the primary (base) goods (from the RFP) offered by the MNEs are homogenous and priced identically. The market structure is such that a prisoner's dilemma-style game could ensue. If all MNEs could coordinate their actions and offer a small offset package, joint profit in the industry would rise. Another scenario is that firm 1 offers a higher-valued offset package than firm 2, *ceteris paribus*, and firm 1 wins the contract. If the firms exhibit similar cost functions, symmetry will exist when firm 2 offers the higher-valued package and wins the contract. Consequently, both (or all in the n player game) firms will offer the higher-valued packages in the Nash equilibrium and earn lower profits. The IAWG, then, seeks to coordinate efforts – perhaps establishing trust among supplier countries – and achieve a cooperative equilibrium. Figure 2.1 depicts the

[15] The U.S. Department of Commerce (2010) *Offsets in Defense Trade* attempts to quantify this substitution effect. Unfortunately, the methodology is flawed and the results cannot be accepted.

simple game. The numbers inside the cells represent the possible monetary payoffs (e.g., in millions) for the firms under each scenario.

Assessing Performance: Metrics

Ex post monitoring and contract evaluation are critical if we are to ascertain the effectiveness of a given transaction. To date, it must be acknowledged that most government attempts to monitor offset performance have been unrefined and imprecise at best, inaccurate and corrupt at worst. How can a government accurately measure performance of this policy instrument? We might start with the two most common metrics in human resource evaluation: inputs and outputs.

Input Metrics

An offset can be written as a function of capital, labor, technology, and other resources: Offset Contract Benefits = (K, L, technology, u_t). The contract calls for a given work or economic benefit to transfer to the purchasing government's economy; u_t is the *portmanteau* variable that includes political, security, and economic development considerations. The valuation of capital and labor is straightforward, as their original costs and current market values are readily available. The value of technology transfer and tacit information is more difficult to quantify. In these markets, asymmetric information and uncertainty create circumstances where buyers know less about product attributes, cost, and future utility than sellers (who themselves are not impervious to uncertainty). These information problems can raise transaction costs, induce opportunism on the part of the seller, and generally distort the evaluation metrics.[16] Legal recourse is unattractive in the international marketplace because enforcement tends to be costly, and outcomes are highly uncertain. Williamson (1983) argues that in the absence of court action, firms employ private-ordered contracts to safeguard the exchange and minimize *ex post* measurement costs.

Consider an offset contract that requires the seller to transfer a specified technology to the purchasing government's economy. In 2007, for example, the Polish military purchased €110 million worth of surface-to-air missiles from the Swedish firm SAAB (Countertrade and Offset 2007). The offset arrangement required SAAB to transfer related technology to select Polish firms. After the original exchange is consummated, safeguards are needed to support offset fulfillment in the presence of moral hazard, asymmetric information, and tacit knowledge. Measurement costs rise if Polish officials use input proxies such as billable hours or number of trainees supervised;

[16] See Taylor (2005) for a transaction cost approach to procurement offsets and countertrade.

officials would still need to determine if the technology transfer met program goals. Furthermore, in professional and creative works, it is difficult to ascertain whether the individual is shirking or on the verge of a brilliant innovation (Fama 1991). Exchange settings marked by high transaction costs, opportunism, and imperfect information are monitored more efficiently by the adoption of output metrics.

Output Metrics

Countertrade offsets are designed to transfer economic benefits to the purchasing government's economy. Theoretically, assessing the performance of work ought to be measured by observation of the final product. One can observe a properly functioning automobile plant, shipping company, or a sugar refining company and their respective outputs. One can compute the economic value added (EVA) by summing the jobs created (and associated wages), and the revenue generated from domestic and export sales. In the aforementioned examples, we can calculate a reasonably precise estimate of net benefits and EVA. Though this method requires more sophisticated accounting and economic analysis, it is probably the best approach to evaluate the performance of a sizeable offset contract.

This method uses accounting, market prices, observable sales, and cost data to verify the success (profit) or failure (losses) of the contract. Of course, the underlying assumption of this approach is that (short run) profits signify relative success of the deal, and losses or negative economic value imply the operation underperformed or failed. Today, several countries monitor offset fulfillment with the EVA approach. The United Arab Emirates (UAE), headed by Dr. Amin El-Din, pioneered its use in the late 1990s.

Ideally, the measurement of performance would include a mix of input and output metrics. Why not focus exclusively on outputs to determine EVA to an economy? For the reason it pays to foster select public–private partnerships, nonprofit organizations, and government provision of public goods. These projects frequently fail the initial expected rate of return and profitability tests, and might be construed as underperforming or failed projects. However, losses and negative benefit–cost ratios are common for new firms in emerging industries. Additionally, lumpy public investments may yield positive profits in the long run – but not the short run – as the firms' average total costs decline, learning-by-doing occurs, and product demand reaches a critical level to exploit scale economies.

Nonprofit, educational, and environmentally based offsets may never register positive profits but could still contribute to the purchasing government's economy in a favorable way. For example, a sale of helicopters (Eurocopter) to Kuwait required the European firm, EADS, to make educational investments and provide training in the country (Countertrade and Offset 2004). Strict adherence to financial variables to determine the longer term value of such a project is problematic. Therefore, exclusive reliance on an output metric – akin to the UAE approach – to judge the effectiveness or performance of an offset is not advised. Likewise, narrow application of input metrics can be misleading because the offset provider has incentive to substitute quantity for quality. This behavior may take the form of perfunctory efforts to merely "get the job done," or the intended use of less skilled workers or the "second team." Suppliers might try to cut costs in the provision of inputs, with little incentive to teach routines and "tricks of the trade" that ensure best practices and long-term success. A linear combination of input and output metrics would minimize the aforementioned shortcomings. A sliding scale could be employed to select the optimal input–output metric for the product and industry in question. For example, if the objective is to develop new export markets for local firms, a counterpurchase offset could be combined with management and marketing training services. These activities can be specified in a contract with input metrics. It is straightforward to compute the value of the countertrade and verify the efficacy of training services with assessment and assurance of learning methods.

If the objective is to foster the transfer of technology and capabilities, a linear combination of input and output metrics minimize transaction costs and the potential for opportunism on the part of the seller. To measure input performance, the purchasing government could track labor hours devoted to training programs for domestic workers, estimate the supply cost of the technology transferred, billable consulting hours, and so forth. Clearly, these input metrics are only weakly tied to overall performance of the offset. In these instances, successful outcomes require the transfer of tacit knowledge. If the government relies solely on input metrics, the seller has incentive to act opportunistically by substituting quantity for quality while guarding core capabilities. Output metrics could help to align the incentive structure of the transacting parties, and promote best practices. In this case, the output metric can include any of the following variables: net income data after a designated period, quantity of output that meets or surpasses a given level of quality control, EVA thresholds in the purchasing government's economy, return on investment, benefit–cost thresholds, and so forth.[17]

Results

The U.S. Commerce Department has compiled offset data since 1993. However, these data are limited to transactions involving American MNEs. Furthermore, American firms are only required to self-report if the value of the offset obligation exceeds $5 million. The database is presented in aggregate form only, which severely restricts its use in microeconomic analysis. Consequently, the US database – while useful in evaluating trends and macroeconomic effects in the defense industry – lacks

[17] Jang and Joung (2008) describe the economic value-added (EVA) method applied to defense procurement and countertrade offsets. The EVA integrates the (1) cost, (2) income, (3) lines of code, and (4) case studies methods to provide an objective and credible offset valuation model.

Table 2.2 Industrial classification of the base product

Industry	% of total
Aerospace	48.51
Automotive	9.90
Communications	4.95
Electronics	4.46
Energy	1.49
Environment	0.50
Financial	0.50
Industrial	1.98
Marine	10.89
Mining	0.50
Munitions	9.41
Specialty metals	0.50
Transportation	6.44

Source for Tables 2.2–2.7: author calculations derived from *Countertrade and offset* archives (2003–2007). The totals may not equal 100 due to rounding

the breadth and transaction-level detail required to study the governance structures of the international market.

To address these shortcomings, a database of international government procurement was constructed. The data were gathered from the archives of the most comprehensive trade journal in the field reporting on countertrade transactions.[18] The database comprises 235 international public procurement transactions from 2003 to 2007. Data for the following variables were entered for each transaction, subject to vetting, and double-blind review to ensure accuracy: *MNE supplier, purchasing country, base product, base good industry, price of the base good, monetary value of the offset, offset type, intended offset project, offset recipient firms, time to completion, penalties*. As is typical of this kind of research, in some instances an observation was discarded due to missing data. While the research is ongoing at this stage, we can report several important findings.

Table 2.2 shows the industrial classification of the base good exchanged. These data confirm anecdotal evidence that countertrade offsets arise in imperfectly competitive markets. The aerospace industry generates the most offset transactions. Among the observations, sales of a base good from the aerospace (defense and civil) industry were responsible for an astounding 48.51% of the total. The marine (10.8%), automotive (9.9%), munitions (9.41%), and transportation (6.44%) industries are also well represented. In the aggregate, these five industries

[18] *Countertrade & Offset* was founded in 1983. It specializes in global intelligence and reporting of structured finance and countertrade contracts worldwide.

Table 2.3 Type of offset contract in relation to the base good

Type	% of total
Direct	52.94
Direct and indirect composite	36.27
Indirect	10.78

accounted for 85% of the contracts that had accompanying offset obligations. These data are consistent with the US government's survey of offset-supplying firms. The results also support the theoretical claim that offsets are only appropriate under highly stylized exchange settings – namely, imperfectly competitive markets with super-normal profits, asymmetric and incomplete information, bounded rationality, and reputational economies (Taylor 2004).

Tables 2.3–2.5 focus on offset fulfillment and should not be viewed in isolation. Table 2.3 classifies the transaction according to offset type. A direct offset requires the seller to provide economic benefits to the purchasing government's country that are related to the base good. This may entail significant coproduction work, maintenance and repairs, or a simple subcontract agreement to supply the tires of an advanced aircraft. Indirect offsets are not related to the base good, and may run the gamut from technology transfer and training, to investment. The third classification is the composite case where a contract calls for both direct and indirect offsets. More than 50% of the transactions called for direct offsets, and another 28% included both direct and indirect work. Somewhat surprisingly, pure indirect offsets lagged far behind. There are economic and methodological reasons for this result. In procurement among industrialized countries, direct offsets are preferable because the purchasing country already has a diversified economy and the countertrade is used to maintain or stimulate the DIB. In this exchange setting, direct offsets can facilitate the acquisition of core and ancillary capabilities more efficiently than indirects.[19] We should also note that the *Direct & Indirect Composite* classification masks the use of indirects. We are unable to determine the extent to which a direct/indirect observation is shared equally, or dominated by a direct or indirect project. This is clearly a limitation of the data.

Tables 2.4 and 2.5 delineate the industry and organizational form of the offset fulfillment. Unsurprisingly, the industries that dominated the export sales of the base good were largely the same industries targeted for offset fulfillment. Offset obligations were most often fulfilled in aerospace (41.74%), automotive (8.7%), transportation (8.7%), marine (6.96%), industrial (6.09%), and munitions (5.22%). These data are consistent with the heavy use of directs, the significant presence of defense contracts in the database, and the general aim to preserve country DIBs whether it be efficient to do so or not. Table 2.5 reports the class of offset fulfillment. These data can shed light on the economic organization of the firms, and the means by which the contract can be safeguarded. Coproduction/local

[19] Furthermore, as discussed earlier, the WTO and EU procurement laws favor direct offsets.

Table 2.4 Industry for fulfillment of the offset contract

Industry	% of total
Aerospace	41.74
Agriculture	0.87
Automotive	8.70
Communications	4.35
Electronics	2.61
Energy	3.48
Environment	0.00
Financial	2.61
Forestry	0.87
Industrial	6.09
Marine	6.96
Medical	2.61
Mining	1.74
Munitions	5.22
Social development	0.87
Software	0.87
Specialty metals	0.87
Tourism	0.87
Transportation	8.70

Table 2.5 Organizational form of the offset fulfillment

Organizational form (class)	% of total
Coproduction/licensed production	30.95
Construction and infrastructure	9.52
Countertrade/buyback/barter	4.76
Finance and investment	9.52
Miscellaneous	0.79
Subcontracting	20.63
Technology transfer	23.81

production/local assembly was the most common class of fulfillment (30.95%), followed by technology transfer (23.81%) and subcontracting (20.63%). Again, these results are consistent with the leveraging of direct offsets, maintenance of the DIB, and capabilities acquisition strategies. Interestingly, buyback and barter (countertrade) agreements accounted for only 4.76% of the total, a notable change from past decades. And while counterpurchase agreements are still prevalent, it is clear that buyback and barter have fallen out of favor with purchasing governments. Historically, buyback and barter arrangements were most commonly used to conserve foreign exchange, or establish a credible commitment (reciprocity) to support the transaction. In the current global economy, conservation of hard currency is not a driving force in international public procurement.

Instead, these contracts can be seen as a means of developing trust and aligning incentives through reciprocal exchange (Taylor 2005). Counterpurchase agreements are also more flexible in that they need not be tied to the base good. It follows, then, that indirect offsets coupled with counterpurchase fulfillment is

Table 2.6 Geographic region of manufacturing supplier

Region	% of total
North America	29.65
South America	0.44
Europe	52.65
Eurasia	6.19
Middle East	7.96
Asia	3.10

Table 2.7 Headquarter country of the manufacturing supplier

Country	% of total
Austria	0.9
Belgium	0.4
Brazil	0.9
Canada	1.3
China	1.8
Finland	2.2
France	17.3
Germany	4.4
Israel	7.5
Italy	6.2
Japan	0.9
Netherlands	0.9
Norway	0.4
Russia	5.8
South Korea	0.4
Spain	4.4
Sweden	4.0
Switzerland	1.3
Turkey	0.4
Ukraine	0.4
United Kingdom	9.7
United States	28.3

more common in developing countries that pursue economic diversification rather than DIB investments.

Tables 2.6 and 2.7 reveal the origins of the offset-supplying firms. In these industries, the markets are dominated by (largely) American and European MNEs. Table 2.6 delineates the region of the supplier (MNE). As expected, Europe (52.7%) and North America (29.6%) account for the lion's share of the export transactions.[20] The discrepancy between Europe and North America can be explained, at least partly, by noting that there are a higher number of European firms in the database. Moreover, the median European firm is smaller (in sales and

[20] This comparison tracks base good transactions, not export revenues.

net income per annum) than the North American firm. If we had tracked supplier export sales revenues instead of transactions, the difference between the two regions would likely narrow. Unfortunately, these data are not available.

Perhaps, most striking is the paucity of transactions from the Asian (3.1%) and South American (0.4%) regions.[21] Clearly, this speaks to the relative market shares of firms competing in these industries. Table 2.7 confirms that most of the successful firms in these oligopoly markets are from the US (28.3%), France (17.3%), and the UK (9.7%). Several characteristics are present in most of these markets. First, industries like aerospace, defense, transportation, and telecommunications are marked by high start-up costs and capital-intensive production processes. Second, natural and legal barriers to entry make it difficult for competitors to join the market. Third, significant economies of scale and learning curve effects exist, thereby putting laggards at a competitive disadvantage. Nevertheless, we should expect a gradual increase in supplier activity from select East Asian countries moving forward. South Korean, Japanese, and Chinese firms are rapidly developing the core capabilities necessary to compete in at least some of these markets. The so-called BRIC countries – Brazil, Russia, India, and China – but particularly Russia (5.8%) will likely become more active in the years to come as their governments continue to implement development strategies that target high value-added industry.

Conclusion

Countertrade offsets are the hybrid offspring of economic, political, and security considerations. In recent years, many countries have used offsets as industrial policy to further economic development objectives. The specific features of the contract reflect these objectives and also the degree of exchange hazard posed by the setting. A procurement policy matrix, then, can illustrate the tradeoffs between markets and hierarchies under different exchange settings, and offer some general guidelines to government officials.[22] It is therefore inaccurate to conclude – as some researchers have – that countertrade offsets are strictly inefficient, or strictly beneficial. Both of these views fail to consider the full menu of policy responses that are appropriate for a given exchange setting. For example, the existence of any of the following variables may alter the exchange setting and thus the policy response: the potential for seller opportunism, the potential for strategic alliances, the level of competition and concentration within the industry, capital–labor ratios, and the existence of external economies and increasing returns.

[21] Interestingly, 8.2% of transactions originated from the Middle East. The Middle East is a net importer of these goods and the associated offsets. The majority of these imports are defense procurement. Israel is largely responsible for the Middle East export transactions in the database.

[22] Taylor (2004) develops a procurement policy matrix that considers transaction costs, capabilities of the firm, and production economies.

In this chapter, we set out to further our understanding of countertrade agreements in international public procurement. The theoretical portion of the chapter revisited the *raison d'être* of these agreements. We showed that under certain circumstances, both buyer and seller can achieve a Pareto-superior outcome by abandoning the price margin of Arrow and Debreu markets. When the transactions costs of using markets are relatively high, firms will explore nonstandard contracts to complete the trade. However, we need to underscore the importance of relying on the high-powered incentives of markets wherever possible. Any deviation from free market exchange should be accompanied by an economic audit that estimates net benefits from a policy's second-best solution.

Countertrade offsets have been shown to support multiple objectives. First, the buyer can leverage its purchasing power to elicit economic benefits from the seller that might otherwise not be forthcoming. Though data are not available, anecdotal evidence suggests that technology transfer is inhibited in free market exchange but not to the same extent in countertrade. Second, countertrade agreements can assist a country in its efforts to penetrate foreign markets, gain reputational capital, learn production methods, and strike new relationships with suppliers (to perform subcontracting, coproduction, and licensed production for example). Third, these arrangements are frequently used to maintain and/or develop domestic DIBs. Despite the fact that it usually is not cost-effective to do so, the purchasing government may elect to produce domestically at a higher cost relative to off-the-shelf purchases, for national security reasons. Fourth, countertrade offsets have evolved to support economic development projects that are unrelated to the base transaction. These indirect offsets are designed to grow local industry that will eventually be competitive in international markets. The UAE's offset program was a pioneer in the use of performance metrics to judge *ex post* results.

A shortcoming in the literature had been a lack of guidance on how government ought to measure performance of the countertrade agreement. In this section we detailed the advantages and disadvantages of several proposed metrics. For the prototypical countertrade contract, we recommend a combination of input and output metrics. Reliance on a single metric – while simpler and more transparent – is usually unable to evaluate contracts designed to serve multiple objectives.

The final section of the chapter presents empirical evidence from a database that tracks countertrade agreements in international procurement from 2003 to 2007. Several important findings can be deduced from the data. First, countertrade transactions are still very much part of the global marketplace. The data suggest that countertrade is as important today as it was a decade ago. The US, EU, and WTO have all, to varying degrees, argued for less reliance on countertrade and offsets in international government procurement. On the other hand, many developing countries view the multifaceted benefits of offsets as an appropriate response to markets traditionally dominated by MNEs from the West. Second, less-developed countries use barter, counterpurchase, and indirect offset arrangements much more than the industrialized West. Conversely, most industrialized countries use offsets that are directly related to the base good. This can be explained by differential technology absorption rates and heterogeneous labor markets.

Furthermore, developed countries often purchase defense and aerospace products to meet at least part of its national security objectives. The direct offsets are used to maintain, and indeed bolster the remaining portions of the domestic defense industry. The extent to which offsets are successful in this regard is unclear.

Third, preliminary analysis of the data reveals that barter and buyback agreements in public procurement have declined considerably compared to previous decades. With respect to variables influencing the selection of countertrade offsets, the conservation of foreign exchange is rejected. Countries seek offsets for a variety of economic and noneconomic reasons: the opportunity to save hard currency is not among them.

References

Arrow K, Debreu G (1954) Existence of an equilibrium for a competitive economy. Econometrica 22:265–290

Baldwin R (1969) The case against infant-industry tariff protection. J Polit Econ 77(3):295–305

Brauer J (2004) Economic aspects of arms trade offsets. In: Brauer J, Dunne P (eds) Arms trade & economic development. Routledge, London, pp 54–65

Chang H-J (2008) Bad Samaritans: the myth of free trade and the secret history of capitalism. Bloosmbury Press, New York

Countertrade and Offset (2003) Special report: Dutch Audit (Issue 3). CTO Data Services Co, London

Countertrade and Offset (2003–2007) Issue archives (Issues 1–24 each year). CTO Data Services Co, London

Countertrade and Offset (2004) Kuwait: an updated account of foreign contractors' offset projects (Issue 2). CTO Data Services Co, London

Countertrade and Offset (2007) SAAB and Poland sign offset (Issue 20). CTO Data Services Co, London

European Defence Agency (2009) Annex A: the code of conduct on offsets. In: U.S. Department of Commerce (ed) Offsets in defense trade, 14th edn. U.S. Government Printing Office, Washington

Fama E (1991) Time, Salary, and Incentive Payoffs in Labor Contracts. Journal of Labor Economics, 9(January):25–44

Hall P, Markowski S (1994) On the normality and abnormality of offset obligations. Defence Peace Econ 5:173–188

Hall P, Markowski S (2004) Mandatory defence offsets: conceptual foundations. In: Brauer J, Dunne P (eds) Arms trade & economic development. Routledge, London, pp 44–53

Hennart JF (1989) The transaction cost rationale for countertrade. J Law Econ Organ 5:127–153

Jang W-J, Joung TY (2008) The defense offset valuation model. DISAM J. [Online]. Available at: http://findarticles.com/p/articles/mi_m0IAJ/is_4_29/ai_n24261651/?tag=content;col1. Accessed 2 Jan 2010

Linarelli J (2003) The WTO transparency agenda: law, economics and international relations theory. In: Arrowsmith S, Trybus M (eds) Public procurement: the continuing revolution. Kluwer Law International, London, pp 235–268

Martin S (ed) (1996) The economics of offsets: defense procurement and countertrade. Harcourt Academic Publishers, Amsterdam

Northrup Grumman (2009) Product listing. [Online]. Available at: http://www.atoz.northropgrumman.com/Automated/AtoZ/L.html. Accessed 10 June 2009

Penrose E (1959) The theory of the growth of the firm. Basil-Blackwell Publishers, Oxford

Reinert E (2007) How rich countries got rich and why poor countries stay poor. Carroll & Graf Publishers, New York

Succar P (1987) The need for industrial policy in LDC's – a restatement of the infant industry argument. Int Econ Rev 28(2):521–534

Taylor TK (2004) Using offsets as an economic development strategy. In: Brauer J, Dunne P (eds) Arms trade & economic development. Routledge, London, pp 30–43

Taylor TK (2005) A transaction cost approach to countertrade and offsets in international government procurement. J Int Business Econ 3(1):117–132

Trionfetti F (2003) Home-biased Government Procurement and International Trade: descriptive statistics, theory, and empirical evidence. In: Arrowsmith S, Trybus M (eds) Public procurement: the continuing revolution. Kluwer Law International, London, pp 223–233

United States Department of Commerce (2007) Offsets in defense trade: annual report to congress, 12th edn. U.S. Government Printing Office, Washington

United States Department of Commerce (2009) Offsets in defense trade: annual report to congress, 14th edn. U.S. Government Printing Office, Washington

United States Department of Commerce (2010) Offsets in Defense Trade: Annual Report to Congress, 15th edn. U.S. Government Printing Office, Washington

United States Presidential Commission on Offsets in International Trade (2001) Status report of the presidential commission on offsets in international trade. United States Presidential Commission on Offsets in International Trade, Washington

Williamson O (1983) Credible commitments: using hostages to support exchange. Am Econ Rev 73:519–540

World Trade Organization (2010) Members and observers. [Online]. Available at: http://www.wto.org/english/thewto_e/whatis_e/tif_e/org6_e.html. Accessed 30 Apr 2010

World Trade Organization (1994) Uruguay Round Agreement on Government Procurement. [Online]. Available at: http://www.wto.org/english/docs_e/legal_e/gpr-94_02_e.htm#articleXVI. Accessed 22 March 2010

Chapter 3
Linking Developmental Deliverables to Public Sector Contracts

Ron Watermeyer

Introduction

ISO 10845-1 (2010) defines procurement as *the process which creates, manages and fulfills contracts relating to the provision of goods, services and engineering and construction works or disposals or any combination thereof.*

Public procurement, because of its nature and size, can have a significant impact on social and economic development (Arrowsmith 1995). A study undertaken for the European Community cites five principle domestic socio-economic or political functions which public sector procurement may be used to achieve, in addition to obtaining the required goods and services. These are (McCrudden 1995):

(a) To stimulate economic activity.
(b) To protect national industry against foreign competition.
(c) To improve the competitiveness of certain industrial sectors.
(d) To remedy regional disparities.
(e) To achieve certain more directly social policy functions such as to foster the creation of jobs, to promote fair labour conditions, to promote the use of local labour, to prohibit discrimination against minority groups, to improve environmental quality, to encourage equality of opportunity between men and women, or to promote the increased utilisation of the disabled in employment.

Governments in both developed and developing countries have responded to the use of procurement to attain policy objectives in a number of ways, ranging from making it mandatory for officials to use procurement to attain socio-economic objectives to ruling out its use for such purposes. Others allow officials' discretion in the use of procurement for such purposes. Certain international trade agreements limit the use of procurement to promote policy objectives by placing prohibitions on discrimination

R. Watermeyer (✉)
Soderlund and Schutte Ltd, Johannesburg, South Africa
e-mail: watermeyer@ssinc.co.za

M.A. Yülek and T.K. Taylor (eds.), *Designing Public Procurement Policy in Developing Countries*, DOI 10.1007/978-1-4614-1442-1_3,

and other restrictive trade measures and/or by rules on contract award procedures, e.g. the European Union rules on procurement and the World Trade Organisation's Agreement on Government Procurement (GPA) (Arrowsmith et al. 2000).

Goods and services are typically procured from contractors beyond the borders of a country when they are not available within a country or when there are problems with the capacity, capability and competitiveness of the local supply bases (Department of Public Enterprises 2007). Governments, in support of national development objectives, are invariably not supportive of being dependent on foreign contractors and would prefer that goods and services be produced by local contractors. The development of a competitive domestic market is a very attractive option as it lowers the costs of services, increases security of supply, provides employment to its citizens and provides a platform for local innovation.

Governments often look to offsets in public sector procurement to encourage local development by means of licensing of technology, investment requirements, counter-trade or similar requirements. Investment and counter-trade deliverables are, however, typically delivered over a period considerably longer than the contract period and are most often unrelated to the contract. This not only distorts markets, but also makes their measurement, evaluation and enforcement difficult.

The alternative is to link developmental deliverables relating to the leveraging of technology, skills transfer and increasing investment in the local industry, to the supply chain within a contract. This can be achieved in a fair, equitable, transparent, competitive and cost-effective manner using most public sector procurement regimes. This approach, however, requires an understanding of procurement systems, strategies and methods for linking deliverables to contracts.

Goals for Procurement Systems

Procurement systems are developed around a set of objectives or goals. These goals may, in turn, be categorised as those relating to good governance (primary goals) and those relating to the use of procurement to promote social and national agendas and sustainable development objectives (secondary or non-commercial objectives) (Watermeyer 2004a, b).

Governments establish their procurement systems and policies either explicitly or implicitly around a set of goals. Such goals may be used as the point of departure for the development of regulations or form part of the legislation itself. For example, the goals associated with the first nation-wide non-binding regulation in China, the Interim Measures for the Administration of Government Procurement (1999), were to unify legislation and to forge an open, just and fair framework for procurement, while the Constitution of the Republic of South Africa (Act 108 of 1996) requires the government procurement system to be fair, equitable, transparent, competitive and cost effective (Watermeyer 2004a, b).

Similarly, secondary objectives regulating procurement may be expressed in law or in policy. For example, the Constitution of the Republic of South Africa

Table 3.1 Basic procurement system requirements (ISO 10845-1 2010)

Attribute	Basic system requirement
Fair	The process of offer and acceptance is conducted impartially without bias and provides participating parties simultaneous and timely access to the same information
	Terms and conditions for performing the work do not unfairly prejudice the interests of the parties
Equitable	The only grounds for not awarding a contract to a tenderer who complies with all requirements are restrictions from doing business with the employer, lack of capability or capacity, legal impediments and conflicts of interest
Transparent	The procurement process and criteria upon which decisions are to be made shall be publicised. Decisions (award and intermediate) are made publicly available together with reasons for those decisions. It is possible to verify that criteria were applied
	The requirements of procurement documents are presented in a clear, unambiguous, comprehensive and understandable manner
Competitive	The system provides for appropriate levels of competition to ensure cost-effective and best value outcomes
Cost-effective	The processes, procedures and methods are standardised with sufficient flexibility to attain best value outcomes in respect of quality, timing and price, and the least resources to effectively manage and control procurement processes
Promotion of other objectives	The system may incorporate measures to promote objectives associated with a secondary procurement policy subject to qualified tenderers not being excluded and deliverables or evaluation criteria being measurable, quantifiable and monitored for compliance

(Act 108 of 1996) establishes South Africa's preferential procurement policy and the Preferential Procurement Policy Framework Act (Act 5 of 2000) establishes the framework within which such a policy is to be implemented. In India, there are policies in place which permit "purchase preferences" (i.e. allowing public sector units who compete with the private sector to revise their prices downwards where their tenders are within 10% of a large private sector unit) and reserve a large number of products for production by small-scale firms and provide for price preference in favour of small-scale firms. In Malaysia, procurement policy supports the National Development Policy which seeks to improve the economic participation of the indigenous people (Bumiputra) and to make them equal partners of development in the country (Watermeyer 2004a, b).

ISO 10845-1 (2010) requires an organisation to *develop and document its procurement system*

(a) In a manner which is fair, equitable, competitive and cost-effective and which may, subject to the policies of an employer and any prevailing legislation, include the promotion of other objectives, in accordance with the requirements of Table 3.1.
(b) Around a process which commences once the need for procurement is identified, ends when the transaction is completed and includes the attainment of procedural milestones which enable the system to be controlled and managed.

Table 3.2 Classification of secondary objectives in terms of obligations placed on contractors

Type		
Number	Descriptor	Obligation placed on tenderer or successful contractor
I	Structure of the contracting entity	Comply with nominated requirements to be eligible for the award of a contract or a score in the evaluation of tender offers
II	Internal workings of the contracting entity	Comply with nominated requirements to be eligible for the award of a contract or a score in the evaluation of tender offers; or undertake to implement certain work place actions during the performance of a particular contract
III	Outsourcing	Provide business or employment opportunities (or both) to target groups through activities directly related to a particular contract
IV	Nominated deliverables	Undertake to provide specific deliverables, which might be related or unrelated to a particular contract

Source: after Watermeyer (2004a)

The promotion of secondary objectives (Table 3.2) is nevertheless required in terms of ISO 10845-1 (2010) to be fair, equitable, transparent, competitive and cost-effective in a measureable, quantifiable and verifiable manner. As such, it may not distort competition or lead to outcomes that are not cost-effective as deliverables are not guaranteed, or are set at a level that most tenderers cannot readily attain in the performance of the contract.

Addressing Secondary Objectives in a Procurement System

Secondary or non-commercial objectives can be categorised in terms of obligations placed on tenderers or successful contractors as tabulated in Table 3.2. The focus of each of the four categories of obligations described in Table 3.2 is different. The Type I and II categories (structure or internal workings of the contracting entity) typically focus on the structure and internal workings within the contractor's organisation immediately before the commencement of the contract and for the duration of the contract. The Type III category (outsourcing), on the other hand, focuses on undertakings that targeted enterprises or labour (or both) will be engaged in economic activities in the performance of the contract, while Type IV (nominated deliverables) uses procurement to leverage socio-economic benefits which are related or unrelated to the contract.

Several models for public sector procurement interventions apart from offsets have evolved, based largely on country-specific procurement regimes and requirements. These can be broadly categorised as falling into one of five generic schemes indicated in Table 3.3, which, in turn, can be subdivided into one of ten implementation methods.

Table 3.3 Methods used to implement policies relating to secondary objective

Scheme type	Methods	Actions associated with the method
Reservation	Set asides	Allow only enterprises that have prescribed characteristics to compete for the contracts or portions thereof, which have been reserved for their exclusive execution
	Qualification criteria	Exclude firms that cannot comply with a specified requirement, or standard relating to the policy objectives from participation in contracts other than those provided for in the law
	Contractual conditions	Make policy objectives a contractual condition, e.g. a fixed percentage of the work shall be subcontracted out to enterprises that have prescribed characteristics, or a joint venture shall be entered into
	Offering back	Offer tenderers who satisfy criteria relating to policy objectives an opportunity to undertake the whole or part of the contract if that tenderer is prepared to match the price and quality of the best tender received
Award criteria	Weighting of objectives at the shortlisting stage	Limit the number of tenderers who are invited to tender on the basis of qualifications and give a weighting to policy objectives along with the usual commercial criteria at the shortlisting stage
	Award criteria (tender evaluation criteria)	Give a weighting to policy objectives along with the usual commercial criteria, such as price and quality, at the award stage
Incentives	Incentive payments	Incentive payments are made to contractors should they achieve a specified target (key performance indicator) in the performance of a contract
Indirect	Product/service specifications	State requirements in product or service specifications, e.g. by specifying labour-based construction methods
	Design of specifications, contract conditions and procurement processes to benefit particular contractors	Design specifications or set contract terms (or both) to facilitate participation by targeted groups of suppliers
Supply side	General assistance	Provide support for targeted groups to compete for business, without giving these parties any favourable treatment in the actual procurement

Source: after Watermeyer (2004a)

Concerns regarding the undermining of primary procurement (good governance) objectives are invariably expressed whenever procurement is used as an instrument of socio-economic policy. Typically, the concerns raised revolve around the risk of the following occurring when implementing a preferential procurement policy (Watermeyer 2004a):

(a) Loss of economy and inefficiency in procurement.
(b) Exclusion of certain eligible tenderers from competing for contracts.
(c) Reduction in competition.
(d) Unfair and inequitable treatment of contractors.
(e) Lack of integrity or fairness.
(f) Lack of transparency in procurement procedures.
(g) Failure to achieve secondary procurement objectives through the procurement itself.

It should be noted that these risks relate to the compromising of a procurement system's good governance objectives.

Type IV (nominated deliverables) secondary objects that are unrelated to a particular contract (offsets) are extremely difficult to implement in a manner which does not violate most of the aforementioned concerns, particularly if obligations extend beyond the procurement contract period. Accordingly, these secondary objectives are considered from the outset to compromise primary procurement objectives.

Watermeyer performed a risk assessment based on AS/NZS 4360:1999 (Risk Management) on the implementation of a preferential procurement policy which has objectives that can be realised by creating a demand for services and supplies from or to secure the participation of, targeted enterprises and targeted labour (Types I and III secondary objectives), using eight of the methods listed in Table 3.3 (Watermeyer 2004a). His analysis indicated that the methods which relate to award criteria at the short listing (method 5) and tender evaluation stage (method 6), which although not guaranteeing that socio-economic objectives will be met, are the methods that are most likely not to compromise requirements for a system which has fair, equitable, transparent, competitive and cost effective good governance goals, if appropriately managed. The analysis furthermore indicated that method 3 (contractual conditions), method 8 (product/description specification), and method 9 (design of procurement to benefit particular contractors) have the potential under certain circumstances to satisfy primary objectives (e.g. where the requirements can be met by most potential tenderers in the performance of the contract), while method 1 (set asides), method 2 (qualification criteria), and method 4 (offering back) are most likely to compromise such objectives. (His analysis omitted financial incentives, which result in the same conclusions as methods 5 and 6.)

Developing a Procurement Strategy

Strategy in the delivery and maintenance of infrastructure may be considered as the skillful planning and managing of the various processes associated therewith. It involves a carefully devised plan of action which needs to be implemented. It is all about taking appropriate decisions in relation to available options and prevailing circumstances in order to achieve optimal outcomes.

Procurement strategy is *the selected packaging, contracting, pricing and targeting strategy and procurement procedure for a particular procurement* (ISO 10845-1 2010). Procurement strategy in the context of the leveraging of technology, skills transfer and increasing investment in the local industry within the supply chain needs to take place at a portfolio level, i.e. where projects and programmes, which are not necessarily interdependent or directly related, are grouped together to facilitate effective management of that work to meet an organisation's strategic business objectives (see Fig. 3.1).

Develop an Infrastructure Plan

Procurement strategy at a portfolio level can only be undertaken after an infrastructure plan which identifies long-term needs and links prioritised needs to a forecasted budget for the next few years has been developed. Such a plan needs to provide a projected list of work items described by category, location, type and function. The work items also need to indicate the nature of the work, i.e. design, construction, installation, refurbishment, supply, rehabilitation and maintenance or any combination thereof.

Identify Developmental Opportunities

The work items in such a plan need to be analysed and further categorised for development purposes in terms of the following (Department of Public Enterprises 2007):

(a) Items for which there is repetitive spend over time, as sustained demand is a prerequisite for the development of local supplier capacity.
(b) High spend items, as this will result in the greatest impact on economic growth.
(c) Items which are currently being imported.
(d) Items for which an increase in expenditure is envisaged.
(e) Items for which there is a lack of competitiveness in the existing supply base.
(f) Items for which there is potential for exports.

The analysis should also consider items that are upstream in the supply chain, i.e. items that are manufactured.

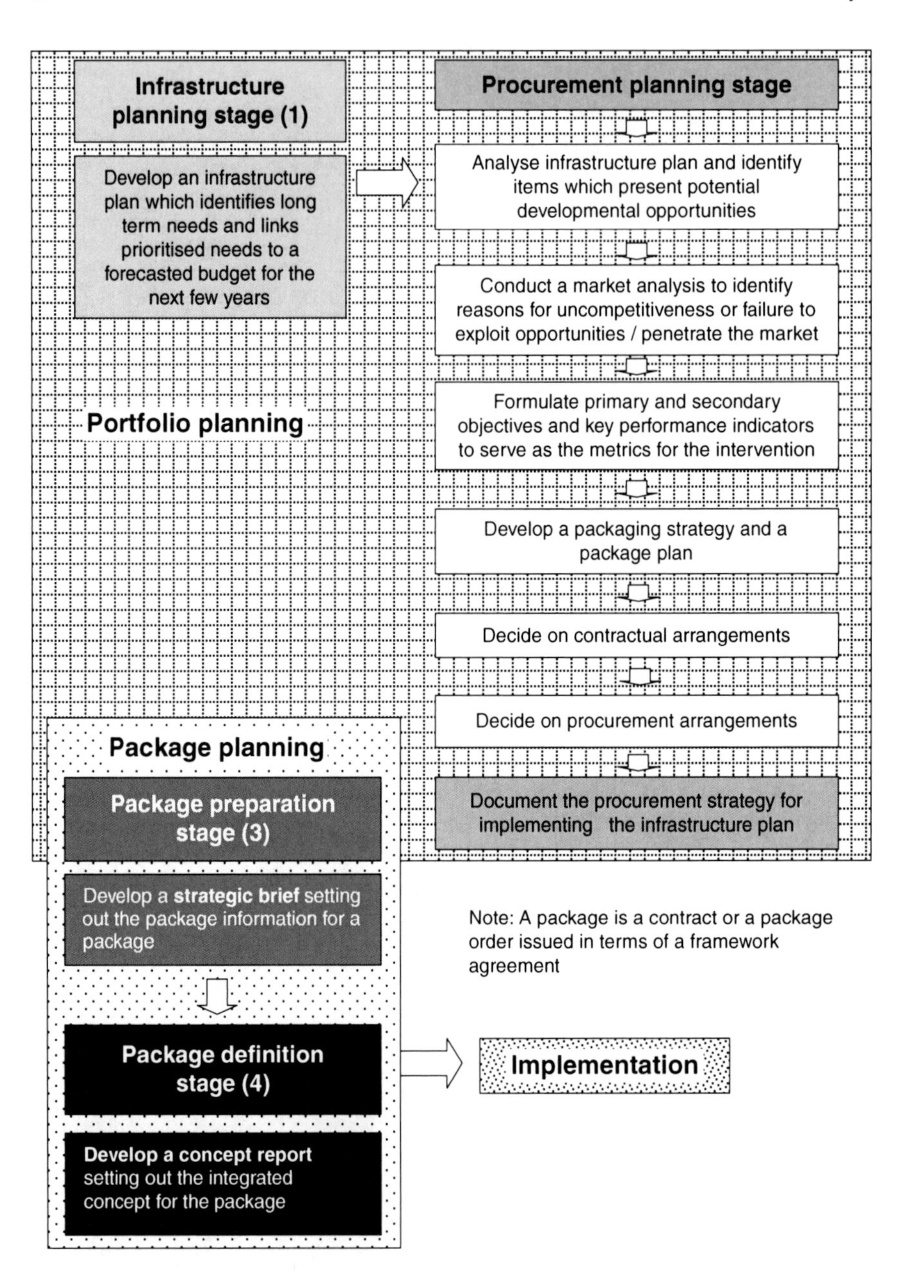

Fig. 3.1 Planning for infrastructure investment at a portfolio and package level

Perform a Market Analysis

The market analysis, which should cover the entire supply chain of all items in the development categories of expenditure contained in the infrastructure plan, needs to identify the causes of uncompetitiveness, insecurity of supply, or the failure of investors to exploit opportunities for expanding local supply. It also needs to identify the local and imported components, and more importantly, the imported items for which there appears to be comparative advantages for local supply and areas in which there is a lack of competitiveness in the local supply base or in which there is a need to increase the security of supply. The characteristics, strengths and weaknesses of the supply networks in these areas should be analysed, including assessing local prices vs. global prices, the cost structure and cost drivers of local suppliers vs. global suppliers, the determinants of competitive advantage, and the technology transfer, skills development and investment which would be required for local production (Department of Public Enterprises 2007).

Research will then need to be carried out to determine the viability of developing the local industry to meet the demand and the potential for export markets. Issues such as the stability of the local and export demand, the competitiveness of local supply vs. global supply, and what demand and supply side interventions need to be considered. The likely impacts also need to be assessed.

Interventions can include increasing the number of suppliers in the market to improve competitiveness or requiring international contractors to form joint ventures with local partners, develop the local industry, invest in local industry, and provide technology transfer and skills development. They can also include changes in designs, standardisation to achieve economies of scale in manufacture, providing the market with information relating to long-term demand, the packaging of contracts, and working closely with industry associations.

Formulate Procurement Objectives and Key Performance Indicators

Primary and secondary procurement objectives are associated with any procurement. Primary procurement objectives typically include:

(a) Tangible objectives including budget, schedule, quality and performance characteristics required from the completed works and rate of delivery.
(b) Environmental objectives.
(c) Health and safety objectives.
(d) Intangible objectives including those relating to buildability, relationships (e.g. long-term relationship to be developed over repeat projects, early contractor involvement, integration of design and construction, etc.), client involvement in the project, end user satisfaction and maintenance and operational responsibilities.

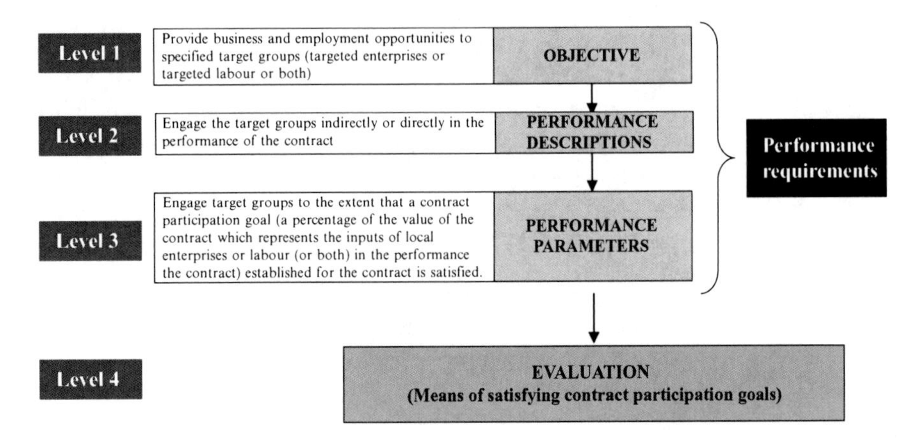

Fig. 3.2 Structure of a four-level performance-based resource specification. *Source*: Watermeyer (2004b) and ISO 10845-5 (2011)

Secondary procurement policy promotes objectives additional to those associated with the immediate objective of the procurement itself (ISO 10845-1 2010). These should focus on objectives relating to local economic development, i.e. the provision of work or business opportunities to local suppliers, contractors and service providers and the development of the local industry to meet the demand. Issues surrounding competing or conflicting primary and secondary objectives will need to be identified and thereafter accepted or resolved.

A key performance indicator (KPI) is a measure of performance which is commonly used to help an organisation define and evaluate how successful it is or to measure progress towards a goal or strategic objective. KPIs should as far as possible be quantitative indicators that can be readily assessed during the performance of the contract.

KPIs relating to the engagement of enterprises, joint venture partners, local resources and local labour in contracts are needed to set targets in contracts or to measure procurement outcomes. Contract participation goals can be used to measure the participation of targeted enterprises or targeted labour, i.e. the flow of money from the contract to the target group. Procedures as to how such goals can be quantified and verified in the performance of the contract should be included in the contract (Watermeyer 2000). This can be done in specifications or other contract information which applies to the contract, as illustrated in Fig. 3.2.

The objective (level 1), the performance description (level 2) and the performance parameters (level 3) as set out in Fig. 3.2 can be viewed as a KPI. The contract participation goal enables targets to be set and evaluation (level 4) establishes the measurement arrangements. A contract participation goal may be used to measure the outcomes of a contract in relation to the engagement of the target groups or to establish a target level of performance for the contractor to achieve or exceed in the performance of a contract (see ISO 10845-5 2011). Quantitative KPIs relating to investments, skills transfer and other contributors to secondary objectives also need to be formulated.

Alternatively, a balanced score card approach can be adopted in terms of which a wide range of KPIs are defined in absolute terms or are qualitatively or quantitatively measured in terms of an indicator. Such indicators need to be objective, verifiable and reproducible, and wherever possible, linked to predetermined benchmarks, reference levels or scales of value which are within levels acceptable to the client. A weighting, which reflects importance, is then assigned to each KPI and the total score measures the performance achieved. This approach enables an overall KPI to be developed.

Develop a Packaging Strategy and Package Plan

A packaging strategy is the organisation of work items into contracts or package orders issued in terms of a framework agreement. (ISO 10845-1 2010 defines a framework agreement as *an agreement between an employer and one or more contractors, the purpose of which is to establish the terms governing contracts to be awarded during a given period, in particular with regard to price and, where appropriate, the quantity envisaged.*) Work items in the infrastructure plan need to be grouped together or divided into packages for delivery under a single contract or a package order issued in terms of a framework agreement.

Framework contracts are well suited to situations in which long-term relationships are entered into. They also offer flexibility in attaining secondary procurement objectives, as requirements can be adjusted from one package order to another, thus allowing new KPIs to be introduced or improved upon over time.

The work items in the infrastructure plan which need to be procured over the next few years need to be grouped or divided into packages by balancing factors such as:

(a) Requirements for independent project/programme of projects.
(b) Use of framework/non-framework agreements.
(c) Geographical spread of project/the technical mix of the work.
(d) Desire to avoid any awkward technical, contractual or logistical interfaces between contracts.
(e) Requirements for management/programme.
(f) Economy of scale from grouping of projects in geographical areas/elimination of duplication of effort.
(g) Marketability, i.e. attractiveness of the packages to the market.
(h) Secondary procurement objectives fit.

A package plan can then be prepared which identifies each package and the timing for the procurement of such packages.

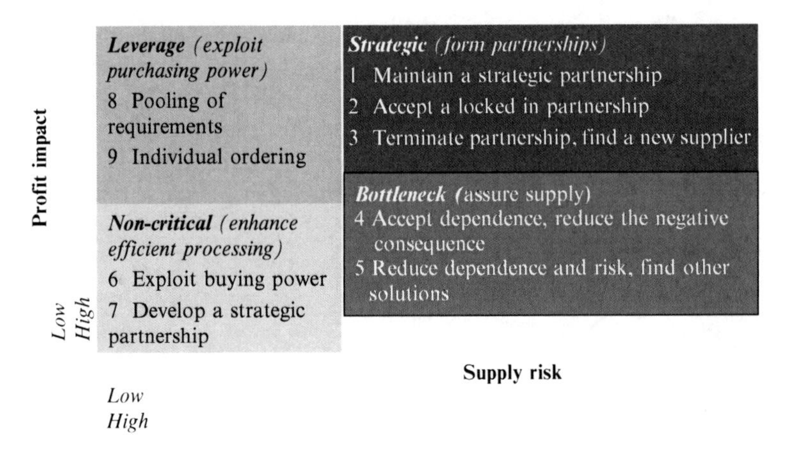

Fig. 3.3 Purchasing strategies within the Kraljic matrix. *Source*: after Caniels and Gelderman (2005)

Decide on Contracting Arrangements

The success or otherwise of development programmes of this nature is dependent upon the decisions that are made when formulating the contractual arrangements. A critical issue to consider is *what culture should be fostered in the contractual relationships*? Alternatively, *what strategy should be adopted?*

Lichtig (2006) has made the observation that design in the construction industry has become increasingly fragmented over the last 100 years and each specialised participant now tends to work in an isolated silo, with no real integration of participant's collective wisdom. Lichtig's proposition is that project success requires that this fragmentation be addressed directly in order to provide higher value and less waste. His research into successful projects has shown that there are several critical keys to success, namely, a knowledgeable, trustworthy, and decisive facility owner/developer, a team with relevant experience and chemistry assembled as early as possible but certainly before 25% of the project design is complete, and a contract that encourages and rewards organisations for behaving as a team.

Kraljic's purchasing portfolio approach (Kraljic 1983) has inspired many researchers to develop purchasing portfolio models following the publication of his matrix in 1983 for single products or product groups (Kraljic 1983). According to Kraljic, a supply strategy is driven by two factors – profit impact and supply risk. Kraljic proposed a four-stage approach for developing supply strategies. First, a company classifies all its purchased products in terms of profit impact and supply risk. Subsequently, it weighs the bargaining power of its suppliers against its own powers. Then, it positions those products that were identified as strategic in a portfolio matrix. Finally, it develops purchasing strategies and action plans for these strategic products, depending on its own strength and the strength of the supply market (see Fig. 2, Caniels and Gelderman (2005)) (see Fig. 3.3).

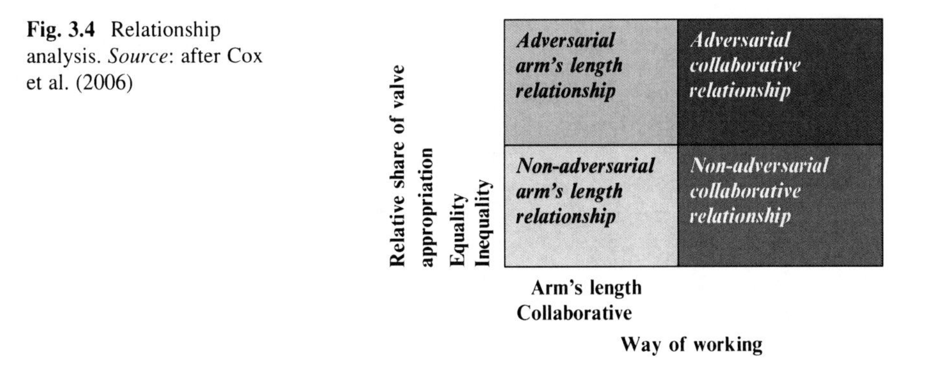

Fig. 3.4 Relationship analysis. *Source*: after Cox et al. (2006)

Cox et al. argue that buyers and suppliers must select from a range of sourcing options and implement them (Cox et al. 2006). They suggest that three elements must be in place, namely the specification of the sourcing approaches, an understanding of the power and leverage environments within which relationships must be managed and an understanding of the relationship management styles that can be used to manage particular sourcing approaches effectively. They argue that all these elements need to be brought together in order to align a particular sourcing approach with a specific power leverage circumstance using the appropriate relationship management style.

The sourcing options link together the level of involvement that buyers and suppliers can have with one another (reactive and arm's length or proactive and collaborative (see Fig. 3.4), as well as the nature and degree of the buyer's involvement in developing supplier and supplier's own competencies) at the first tier or throughout the supply chain(s) as a whole (see Fig. 3.5). Supplier development and supply chain management sourcing approaches are only really effective in situations of buyer dominance and independence (see Fig. 3.6). Buyers will, however, have to normally adopt reactive sourcing approaches where this is not the case because they will be unable to provide the necessary incentives to induce suppliers to invest in the dedicated investments and relationship-specific adaptations to make a proactive approach possible. Cox et al. suggest (Cox et al. 2006) that successful performance outcomes for buyers and sellers require an alignment between goals and aspirations to make the relationship successful for both parties.

The Institution of Civil Engineers has indicated (Institution of Civil Engineers 2010) that a cultural change along the lines of that outlined in Table 3.4 is required to improve and optimise the delivery of infrastructure. They also make the observation that the choice of the contracting system can facilitate or frustrate performance.

The choice of the contracting system for a development programme can facilitate or frustrate performance in terms of the required project outcomes. Ideally, the selected form of contract should enable equitable long-term collaborative relationships to be entered into, be sufficiently flexible to enable risks to be effectively managed, reward performance and focus the parties on attaining the project

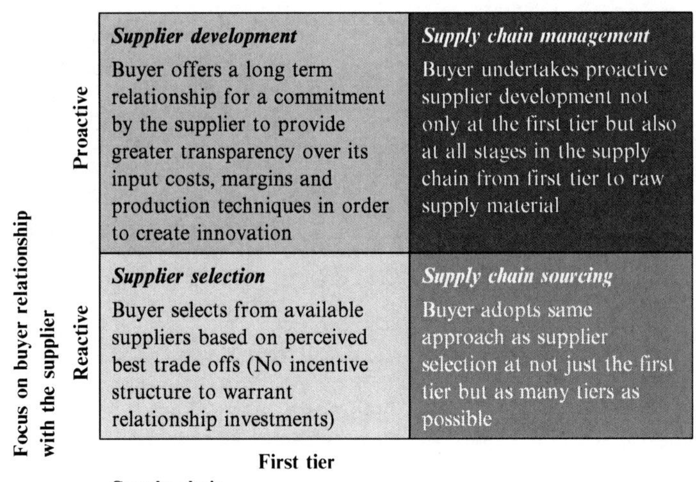

Fig. 3.5 Sourcing options. *Source*: after Cox et al. (2006)

Attributes of supplier relative to buyer		
High	***Buyer dominance*** Few buyers and many suppliers Buyer's account is attractive to supplier Supplier's offerings is a standardised commodity	***Interdependence*** Few buyers and suppliers Buyer's account is attractive to supplier Supplier's offerings are relatively unique
Low	***Independence*** Many buyers and suppliers Supplier has little dependence on buyer for revenue and has many alternatives Supplier's offerings is a standardised commodity	***Supplier dominance*** Many buyers and few suppliers Supplier has no dependence on buyer for revenue and has many alternatives Supplier's offerings are relatively unique

Attributes of supplier power relative to buyer

Fig. 3.6 Power leverage analysis. *Source*: after Cox et al. (2006)

outcomes throughout the supply chain. The system should also cover the full range of contract types, namely supply, services and works, and provide back-to-back subcontracts.

The NEC3 family of contracts, published by the Institution of Civil Engineers, is a set of standard forms of contracts which not only define the legal relationship between the parties to a contract, but also facilitate the implementation of sound project management principles and practices. These contracts, which are suitable

Table 3.4 The culture shift that is necessary to improve and optimise the delivery of infrastructure

From	To
Master–servant relationship of adversity	Collaboration between two experts
Fragmentation of design and construction	Integration of design and construction
Allowing risks to take their course or extreme and inappropriate risk avoidance or risk transfer	Active, collaborative risk management and mitigation
Meetings focused on the past – what has been done, who is responsible, claims, etc.	Meetings focused on "How can we finish project within time and available budget"?
Develop project in response to a stakeholder wish list	Deliver the optimal project within the available budget
"Pay as you go" delivery culture	Discipline of continuous budget control
Constructability and cost model determined by design team and cost consultant only	Constructability and cost model developed with contractor's insights
Short-term *hit-and-run* relationships focused on one-sided gain	Long-term relationships focused on maximising efficiency and shared value
Procurement strategy focused on selection of form of contract	Selected packaging, contracting, pricing, and targeting strategy and procurement procedure aligned with project objectives
Project management focused on contract administration	Decisions converge on the achievement of the client's objectives
Training is in classrooms unconnected with work experience	Capability building is integrated within infrastructure delivery
Lowest initial cost	Best value over life cycle

Source: Institution of Civil Engineers (2010)

for use across the full spectrum of works, services and supply contracts ranging from major framework contracts or major projects to minor works contracts or the purchasing of readily available goods, are designed to encourage collaboration and teamwork in contributing to and delivering best value outcomes. They are sufficiently flexible to be used across the entire supply chain in the delivery and maintenance of infrastructure.

The NEC3 Engineering and Construction Contract, NEC3 Engineering and Construction Subcontract, the NEC3 Term Service Contract and the NEC3 Professional Service Contract all have a target contract option. (ISO 6707-2 (1993) defines a target cost contract as *a cost reimbursement contract in which a preliminary target cost is estimated, and on completion of the work, the difference between the target cost and the actual cost is apportioned between the client and the contractor on an agreed basis.*) This contracting arrangement not only enables framework contracts to be entered into, but also enables the client to know where the money is being spent, rewards strong contractor performance, shares financial risk between the client and the contractor and promotes collaboration or a culture whereby both parties have a direct interest in decisions that are made regarding the cost and timing of the contract.

These NEC3 contracts, as well as the NEC3 Supply Contract, have standard options for partnering (option X12) which requires each partner to work with the other partners to achieve the client's objectives and links financial incentives to achieving or exceeding the target stated for a KPI. This option enables a project team to manage their performance with respect to KPIs and to introduce new KPIs as the need arises. As an alternative, each of these contracts has an option (option X20) which allows for incentive payments to be made if a target for a KPI is achieved or exceeded. The NEC3 family of contracts is accordingly well placed to support the required culture change and broader project objectives required to support development projects.

Decide on Procurement Arrangements

Decisions need to be made regarding procurement methodologies. This necessitates that decisions be made regarding:

(a) How quality is to be achieved (see ISO 10845-1 2010), e.g. through specifications, tender evaluation criteria, prequalification, undertakings during the tender process, etc.
(b) The selection of a tender evaluation method and a procurement procedure (see Fig. 3.7).
(c) Targeted procurement strategies to promote secondary objectives.

There are a number of techniques and mechanisms associated with targeted procurement procedures, (Watermeyer 2000, 2004a, b; ISO 10845-1 2010) all of which are designed to promote or attain the participation of targeted enterprises and targeted labour in contracts. These procedures relate to the:

(a) Measurement and quantification of the participation of targeted groups through monetary transactions with such groups.
(b) Definition and identification of targeted groups in a contractually enforceable manner.
(c) Unbundling of contracts either directly so that targeted enterprises can perform the contracts as main contractors or indirectly through resource specifications which require main contractors to engage target groups as subcontractors, service providers or suppliers within the supply chain or as joint venture partners.
(d) Granting of evaluation points in the evaluation of expressions of interest or tenders (preferences) should respondents or tenderers satisfy specific criteria or undertake to achieve certain goals or KPIs in the performance of the contract.
(e) Provision of financial incentives for the attainment of KPIs in the performance of the contract.
(f) Creation of contractual obligations to engage target groups in the performance of the contract, e.g. subcontract a percentage of the work to or contract goods or services from targeted enterprises, enter into joint venture with targeted enterprises,

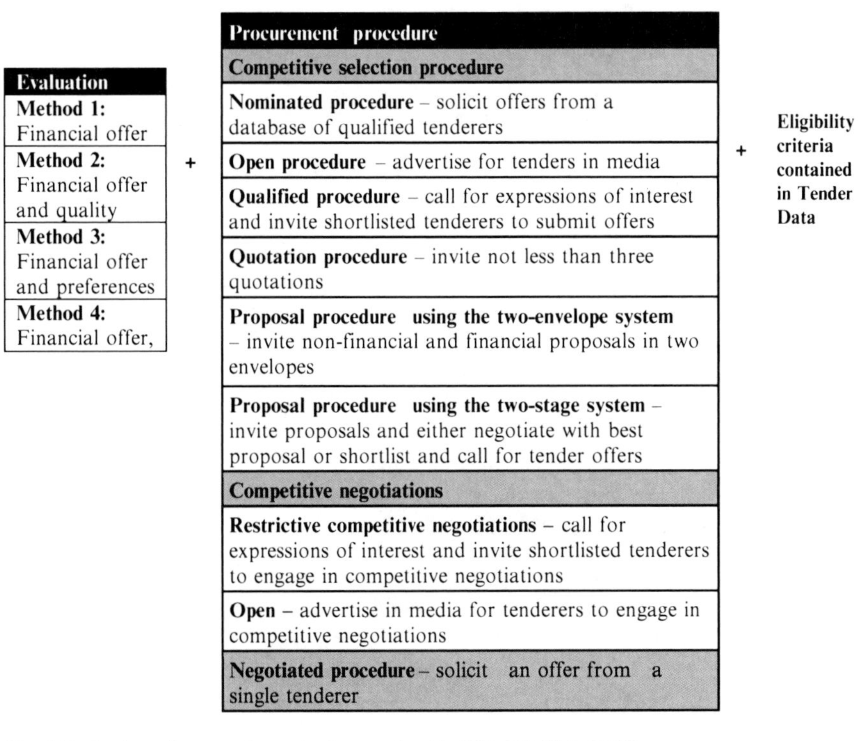

Fig. 3.7 Options for generic methods contained in ISO 10845-1 (2010)

subcontract specific portions of a contract to targeted enterprises in terms of a specified procedure or perform the works in a manner such that targeted labour is employed.

(g) Evaluation of procurement outcomes, i.e. the monitoring of the attainment of socio-economic deliverables at a contract level.

Conclusion

Public procurement, because of its nature and size, can have a significant impact on a country's social and economic development agenda. Developmental deliverables can be linked to the provision of goods and services in terms of the contract. This can be achieved in a fair, equitable, transparent, competitive and cost-effective manner using most public sector procurement regimes. This, however, requires a thorough understanding of the options that are available and the development of appropriate procurement strategies at a portfolio level to exploit the development opportunities that can be linked to procurement. Targeted procurement procedures, framework agreements, term contracts, partnering and target cost contracts have a role to play in this regard.

References

Arrowsmith S (1995) Public procurement as an instrument of policy and the implementation of market liberalisation. Law Q Rev 1995:235–284

Arrowsmith S, Meyer G, Trybus M (2000) Non-commercial factors in public procurement. The Public Procurement Research Group, School of Law, Nottingham University, Nottingham

Caniels M, Gelderman CJ (2005) Purchasing strategies in the Kraljic matrix – a power and dependence perspective. J Purchasing Suppl Manage 11:141–155

Cox A, Ireland P, Townsend M (2006) Managing in construction supply chains and markets. Thomas Telford, London

Department of Public Enterprises (2007) Introduction to the competitive supplier development programme. Republic of South Africa, http://www.dpe.gov.za/res/DPE1.pdf. Accessed 1st Oct 2011

Institution of Civil Engineers (2010) Accelerating infrastructure delivery–improving the quality of 440 life. Second ICE Middle East and Africa Conference, http://www.ice-sa.org.za. Accessed 15 Jan 2010

ISO 6707-2 (1993) Building and civil engineering-Vocabulary – Part 2: Contract Terms

ISO 10845-1 (2010) Construction procurement – Part 1: Processes, methods and procedures

ISO 10845-5 (2011) Construction procurement – Part 5: Participation of Targeted Enterprises in Contracts

Kraljic P (1983) Purchasing must become supply management. Harv Bus Rev 61(5):109–117

Lichtig WA (2006) The integrated agreement for lean project delivery. Construction Lawyer, Summer pp 25

McCrudden C (1995) Public Procurement and Equal Opportunities in the European Community. A study of contract compliance in member states of the European Community and under European Community Law, Contract File No. SOC 9310257105B04, Oxford University

Watermeyer RB (2000) The use of targeted procurement as an instrument of poverty alleviation and job creation in infrastructure projects. Pub Procurement Law Rev 5:201–266

Watermeyer RB (2004a) Facilitating sustainable development through public and donor regimes: tools and techniques. Pub Procurement Law Rev 1:30–55

Watermeyer RB (2004b) Project Synthesis Report: Unpacking Transparency in Government Procurement – Rethinking WTO Government Procurement Agreements. In: CUTS Centre for International Trade, Economics and Environment, Unpacking Transparency in Government Procurement. CUTS International, Jaipur, India, pp 1–50

Chapter 4
Industrialization Strategy Based on Import Substitution Trade Policy

Houssam-Eddine Bessam, Rainer Gadow, and Ulli Arnold

Introduction

Industrialization is the main hope for most poor countries trying to increase their levels of income. Industrialization through import substitution (ISI) is one among many industrialization strategies. It is a relatively old strategy which was applied early during the industrialization of Western Europe. It is based on the protectionism of the domestic production capacity. It was later applied by Latin American nations since the 1930s to speed up their industrialization process. The theoretical bases of the ISI were first developed by Raul Prebisch and Hans Singer in 1950 who suggested the procurement of production means to decrease the imports of manufactured goods in order to cope with the phenomenon of terms of trade deterioration of the LDCs. Several authors have stressed the ISI strategy aspects such as its economic measures (like taxation), its historical aspects and also the great case studies on it around the world in Latin America and south-east Asia. However, ISI was not studied at a microlevel, which is the company level, and nobody asked the question how one should operate a company in a country where the industrial strategy is ISI-oriented.

This paper is a first attempt to build a framework for an international technology transfer (ITT) project. The suggested framework is seen from the point of view of the company interested in carrying out an ITT project. The company is represented as a proactive participant in the whole process.

The same framework could be used by the government to define the national industrial strategy in order to make it concentrated on the companies' needs.

H.-E. Bessam (✉) • R. Gadow • U. Arnold
University of Stuttgart, Stuttgart, Germany
e-mail: houssam-eddine.bessam@gsame.uni-stuttgart.de

M.A. Yülek and T.K. Taylor (eds.), *Designing Public Procurement Policy in Developing Countries*, DOI 10.1007/978-1-4614-1442-1_4,

The paper is composed of three main parts:

The first part is a literature review of the ISI strategy consisting of the localization of the ISI among other development strategies, the ISI theoretical aspects and case studies about ISI to conclude the paper.

The second part is devoted to the development of the framework description through its main stages.

The third part is a simple application of the developed framework on a selected case and, finally, the conclusions.

Underdevelopment and Its Symptoms

According to Gélinas (1994), underdevelopment can be defined from its symptoms: such as malnutrition, infant mortality, illiteracy, GNP (gross national product) per capita and debt; underdevelopment is characterized by:

- Financial dependence and monetary policy: Debt (principal and interest) is so important that the country must rely on external financial markets to repay. Its economy is so extremely sensitive to any "sudden fluctuations in interest rates and exchange."
- The extroversion of the economic system: The economy of an underdeveloped country is based mainly on the export of raw materials with little added value. They do not control the prices of manufactured goods or the food they import. They have no effective way to change fairly the terms of trade in their favor.
- Financial dependence and commercial extroversion generate and maintain a dual society. On the one hand there is the export sector, "forced to adapt its products, its technology, and management to external conditions," and on the other hand the traditional sectors.
- The subordination of elites to outside interests: The mechanisms of development have given rise to a class of politicians, technocrats, and bureaucrats connected to international aid.

Development Strategies

Economic development has proceeded in different ways and through different systems or policies whose main objectives are to stimulate the economy.

Since the 1950s, the used means were evolved to be adapted to the socio-economic context of each era and situation. Some of these strategic choices facing the developing countries are:

- Industrialization or agriculture?
- Protectionism or trade openness?

- Planning state or private enterprise?
- Where to find the capital?
- What technology to use?

Development Through Agriculture

Data from 2002 show that

- In less developed countries, agriculture represents 69% of the workforce and 33% of GDP (the World Bank [WB]).
- According to the FAO, hunger remains a reality in that 17% of the populations of the developing countries (815 million people) suffer from malnutrition, while five million children die each year.

However, agriculture is neglected in many LDCs' strategies. Several authors have shown that food aid is not a solution for malnutrition and underproduction of agriculture. Historically, it is proven that it is not possible to realize industrial development if the agriculture is not developed enough:

- The European Industrial Revolution was preceded by an agricultural revolution (see Bairoch 1973)
- Industrialization of Taiwan, South Korea, and China has been preceded by an agricultural revolution (Gélinas 2001)

Agricultural development has mainly two tools (Gélinas 2001):

- Land reforms: Right to land is unevenly distributed in the Third World, particularly in Latin America for e.g., in Brazil: 1% owns 46% of land.

Inequality in the distribution of land causes:

- Underemployment and exploitation of workers and agricultural land
- Rural Exodus

Land reform aimed at correcting this situation by redistributing the land of large landowners:

- Redistribution in individual plots e.g., Korea and Taiwan (50s)
- Transfer to production cooperatives e.g., Mexico (1917) and China (1958)

Agricultural modernization and the green revolution aim to increase agricultural productivity to achieve food self-sufficiency and/or develop export agriculture.

- Agricultural modernization: the development of chemical fertilizers and pesticides, irrigation, optimization techniques, and mechanization
- Green Revolution: development since the 60s, varieties of high yielding plants obtained by agricultural research:

- Mexico: Wheat and maize
- Philippines, India, Pakistan, Indonesia: Rice
- Africa which was initially untouched by the revolution

Development Through Industrialization and Modernization

Industrialization of the developing countries could be achieved through a variety of strategies. Some of them are presented in the following paragraphs. However, the emphasis is placed on ISI because it is the subject of this paper.

Balanced and Unbalanced Growth

Balanced approach to growth e.g., Nurkse (1961) sets out a strategy in which a country must simultaneously launch a broad range of industries to achieve sustained growth. It aims to create enough jobs to generate demand for new industrial products. Therefore a large initial investment is required. A "big push" necessitates foreign investment (private investment and international aid) and calls for an important role for the state for funding and planning (Nath 1962).

Under *approaches to unbalanced growth forward and backward linkages* (A.O. Hirschman) (Hoshino 2001) relate to the fact that growth usually begins in some sectors and regions and create a succession of imbalances. There exist between various branches of industry links to exploit backward linkages so that the establishment of an industry will create demand for inputs. For example, the automotive industry can become a factor of production for another industry. For example, oil drilling encourages establishment of refineries.

The Industrializing Industries

This approach was developed by Perroux and Bernis, (Destanne de Bernis 1974) and its objective is to develop certain industries that have a strong ripple effect on the rest of the economy. As a result, the priority is given to heavy industries aiming to take advantage of forward linkages. It is characterized by expensive investments that impose heavy state intervention for planning and nationalization of enterprises.

Export-Oriented Industrialization

The objective is to tap the vast market of the industrialized countries. It is a strategy in which a country would gradually replace the traditional exports with the export of manufactures (Karunaratne 1980). Currently it is the dominant strategy advocated by the International Monetary Fund (IMF), and the WB.

Examples of application:

- Asian Tigers (Korea, Taiwan, Hong Kong, Singapore)
- Mauritius
- Zones: Mexico, China, Southeast Asia

Applied Measures

- Maintaining a rather weak currency to promote export (Karunaratne 1980).
- Supporting export industries by grants and loans at favorable interest rates (Karunaratne 1980).
- Establishing, in several countries, "free trade zones" to attract multinationals that want to produce for export (Karunaratne 1980).
- Using cheap labor to export low value-added products (e.g., textiles).
- "Chain climb" switching to the production of intermediate goods and machinery (Destanne de Bernis 1974)).

Development Through Debt and International Cooperation

One of the measures of development through debt and international cooperation is Debt-for-Development or Debt swap which is shorthand for a transaction in which a government or organization in a creditor country retires a fraction of a developing country's external debt, in exchange for a commitment by the debtor government to invest local currency in "designated programs" (Rosen et al. 1999). In essence, the debt swap concept can also be viewed as a form of foreign assistance to a debtor country by its creditors, whether these creditors are commercial institutions or simply credit granting governments.

Development Through Democratization

It is revealed that the institutional and legal development that lays the path for a stable democracy has a beneficial impact on economic growth (Boix 2009). As such, the role of international assistance and nongovernmental entities in strengthening the legal and institutional framework for democracy could increase the prospects for positive economic development in the underdeveloped world.

Import Substituting Industrialization (ISI)

The objective is to produce locally the previously imported goods, which means substituting the imports of manufactured goods by imports of equipment and machinery (Felix 1965). It was recommended by the Economic Commission for Latin America & Caribbean (ECLAC) in the 1950s (Werner 1975) and was the dominant strategy in the Third World during the 1960s (Waterbury 1999). In relation to the debate on "protectionism versus trade openness," ISI strategy is based on the protection of local infant industries from international competition in order to enable them to acquire experience (Mazumdar 1991).

ISI was used by virtually all industrialized countries in the early stages of their development. The Korean economist Ha-Joon Chang, in his book *Kicking Away the Ladder*, (Chang 2002) argues, based on the economic history, that all major developed countries – including the United Kingdom – used interventionist economic policies to promote industrialization and protect national companies until they had reached a level of development in which they were able to compete in the global market. After a period of protection, those countries adopted free market discourses directed at other countries in order to obtain two objectives: (a) to open their own markets to local products and (b) to prevent the newcomers from adopting the same development strategies which created the industrialized nations of today.

Applied Measures

- *Export subsidy*: A fiscal incentive promoting the development of export industries at home, arguably to unfair advantage against the international firms.
- *Foreign exchange controls*: Government measures to ration foreign exchange in order to restrict the quantity of imports or to direct imports to certain sectors. This generally involves compelling exporters to sell foreign exchange to the government at a fixed price.
- Selective importers of key goods are offered preferential rates for foreign exchange, whereas importers of luxury items pay more local currency for their hard currency.
- *Import licensing*: The legal requirement to obtain a license to import certain kinds of goods.
- *Industrial incentives*: Direct payments or tax breaks provided to a firm engaged in a particular line of production.

- *Quota*: A quantitative limit on imports. It places a fixed limit on the quantity of goods that may be imported.
- *Tariff*: The most common type of protectionism in the form of a tax on imports. A tariff works best when the demand for the good in question is elastic or price sensitive. If buyers do not respond to the higher price, a tariff will not limit imports. With a tariff, the central government collects revenues (Chenery 1989).
- Temporary protection.
- First investment target: Consumer goods sector (e.g., clothing, drink, food, shoes).

Results

- High growth rate during the first phase of industrialization (e.g., Brazil, Mexico, Korea, Taiwan)
- Many problems (e.g., Latin America, India):
  - Lack of competition → low productivity firms
  - Businesses repeatedly returning to the state for additional aid (rent seeking)
  - Protection often becoming permanent
  - Small domestic market, limiting growth
  - Dependence on imported capital goods
  - High exchange rate → lower exports and foreign currency
  - Budgets of some states (especially in Africa) becoming dependent on the collection of tariffs

Singer-Prebisch Thesis

As a development policy, ISI is theoretically based on the Singer-Prebisch thesis (SPT) (Singer 1998), which states that countries that export primary products (like most LDCs) have to import less and less for a given level of export.

Singer and Prebisch examined data over a long period of time and they noticed that the terms of trade of the raw materials exporting countries have deteriorated against of the manufactured goods exporting countries since 1876. According to Prebisch, the process of terms of trade deterioration was due to differences in specialization between "northern states" that were technological, and "southern states" whose economies were based on the exploitation of primary resources because:

- The northern states produced manufactured products whose technology and prices increased continuously.
- Southern states provided to the North raw materials whose prices have declined steadily.

For the southern states not yet industrialized, the gradual deterioration in the terms of trade corresponds to a reduction in the purchasing power nationally in terms of foreign goods. Indeed, for the same quantity of raw materials produced and sold to "northern states," they are able to buy a much more reduced quantity of manufactured products.

Prebisch argued, inter alia, that for this reason, the least developed countries should diversify their economies and reduce their dependence on exports by developing their own manufacturing facilities. For Singer, this thesis has joined the "mainstream" theories, because the UN economists were using it in their recommendations of care, made to the countries exporting agricultural commodities. Namely, if prices go up, they should distrust the Dutch Disease; the increase in currency is temporary until a future fallout occurs.

The ISI Case Studies

The ISI in Latin America

The ISI in Mexico

The period from 1930 to 1970, nicknamed as the "Mexican miracle" by economic historians, was a period of economic growth fostered by a model of ISI, which protected and promoted the development of domestic industries. Thanks to ISI model, the country experienced an economic boom during which industries increased production rapidly (Meynier 1953). Major changes in the economic structure included the free distribution of land to farmers subject to the concept of the Ejido, the nationalization of oil and railways, the introduction of social legislation in the constitution, the emergence of labor unions affecting and improving infrastructure. While the population has doubled from 1940 to 1970, the GDP has increased by six times. The ISI model had reached its peak in the late 1960s. ISI in Mexico was a result of favorable domestic demand conditions and temporary foreign supply shortages. The commercial market in Mexico has influenced the pattern of industrialization and the current structure of production, but the demand conditions constituted the leading elements in this process. In the first stage, it was mainly an increase in the production of nondurable consumer goods, gradually shifting toward durable consumer goods, intermediate goods, and on a smaller scale, capital goods (UN 2003).

The deliberate participation of the authorities providing orientation, protection, and sometimes financial resources, accelerated the process of ISI during the 1960s even in the area of capital goods. However, protection that was intended to be temporary has become permanent and the ISI process has become more difficult as further industrialization required pursuing import substitution in new lines of production and increasing efforts to expand industrial goods exports (Llu 1969).

Import substitution has been an integral part of Mexico's development strategy, but from the 1960s it has had disappointing results. Two particularly persistent problems have been the geographic concentration of the new ISI industries and their capital-intensive nature. Such a high degree of capital-intensity has caused the new industries to have little effect on Mexico's chronic unemployment problem. These problems, coupled with the tendency of ISI to produce inefficient plants which require continued protection, have led to a search for alternative policies. One such policy was the export promotion scheme which appeared in the 1960s. Such programs appear to have the greatest promise for Mexico's future (Aboites 1987).

ISI in Brazil

The idea was introduced from the 1930s. It was influenced by a concept of development "modernizing" of the ECLAC of the UN in the postwar context, and according to it, development is seen as a process that is to be undertaken step by step (Auroi 2009).

ISI process began in the 30s by the substitution of imports of nondurable goods such as medicines. Then, production deepened in the 60s, with consumer durables such as automobiles. During the years 1970–1980, the production of intermediate goods and capital goods, such as petrochemicals production and manufacture of turbines for nuclear plants, was the order of the day (Fauré and André 2003). If the ways of applying the model varied greatly during these periods, the "developmentalist" project, which was applied from the 30s under the auspices of Getúlio Vargas, was still present. It consisted of several innovations, including: the creation of public enterprises, the mobilization of public financing, sectoral investment policy, the affirmation of human labor, etc. (Rollinat 2005).

Between 1951 and 1954, during the second government of Vargas, a strategy of developing a large and well-defined plan was adopted. It started with a clean monetary and financial consolidation, and then placed an emphasis on growth in the infrastructure sector, especially in ports, transport, and energy through an influx of foreign capital. The creation of the National Economic Development Bank (BNDE, later becoming BNDES for its interventions in the social field), is one of the important achievements of this period (Samuel 1978).

In conclusion, applying the model of the ISI, Brazil has set up development strategies, whose heritage has long been present in the Brazilian society, as the regime of the CSW (Consolidation of the Statutes of Workers) Vargas era – The government and the IMF considered overly protective of the employee and the source of obstacles in the context of liberalization of the economy. This model depends on the "developmentalist" ideology of ECLAC, which was in crisis in the 60s. While the ECLAC began at that time emphasizing the need for structural reforms in the development process, it was already too late because the military dictatorships were emerging (in 1964 in Brazil) and the instigators of the model were forced into exile (Auroi 2009).

However, the knockout for the ISI model would come not from difficulties of countries in the *periphery* such as Brazil, but from the changes at the *center* which will lead to a new *transnational order*. Enjoying rapid economic growth, the expanding multinationals were to start looking for new markets and cheaper production locations for their manufacturing processes that were becoming more and more technological at the same time (Auroi 2009).

The ISI in Argentina

Argentina began its industrialization in the early 1930s when the depression deprived it of some of its traditional export markets, causing the collapse of its foreign exchange resources, and thus, its import capacity. Many industries, both domestic and foreign were implanted. They began their activities in the field of consumer nondurables and semi-durables (food, clothing, textiles) and then extended them to consumer durables (furniture, appliances, cars) and in late 1960s to capital goods and to a lesser extent, intermediate products (Martin 1967).

Chronologically it is estimated that the years 1948–1950 was the point of cleavage between the two stages (FMCG and more sophisticated products) and the beginning of the industrial block. This was the result of the interplay of exhaustion of easy substitution (simple technology and low capital intensity) and the increasing difficulties to import capital goods. This external bottleneck was itself the result of the product stagnation (in value) of exports with the growth capital needed to continue the process of substituting among branches more and more "capital-intensive" (De Pablo 1977).

Lessons Learned from Latin American ISI

The many forward and backward linkages of automobile manufacturing prompted Latin American governments to give it a central role in their development strategies. So the automobile industry is one of the most important sectors in many Latin American economies. In Argentina, the industry and its linkages account for 22% of employment. After oil, automobiles and automobile components constitutes the second most important export in Mexico. Brazil and Argentina combined are expected to surpass Germany in automobile sales by the first decade of the twenty-first century. We may be surprised to learn that the automobile industry in Latin America is not merely an assembly operation. Full-fledged production has been in place since the early 1950s; Brazil and Mexico have become centers of innovative production. As in the developed countries, the automobile industry occupies a central role in many Latin American economies, although it has not brought the same level of development. The industry's success today dates to the ISI policies of the 1950s and 1960s and the export policies of the 1970s.

The ISI model's openness to other economies in the region is almost zero. Another defect of the model comes from the fact that there is little economic

Table 4.1 Foreign share of selected industries, circa 1970 (percentage)

	Argentina	Brazil	Chile	Colombia	Mexico	Peru	Venezuela
Food	15.3	42.1	23.2	22.0	21.5	33.1	10.1
Textiles	14.2	34.2	22.9	61.9	15.3	39.7	12.9
Chemicals	34.9	49.0	61.9	66.9	50.7	66.7	16.5
Transport equipment	44.4	88.2	64.5	79.7	64.0	72.9	31.1
Electrical machinery	27.6	83.7	48.6	67.2	50.1	60.7	23.2
Paper	25.7	22.3	7.9	79.3	32.9	64.8	20.1
All manufacturing	23.8	50.1	29.9	43.3	34.9	44.0	13.8

Source: Jenkins (1984)

complementarity between the countries of the region. Thus, in the 60s, the failure of the ISI model becomes evident in its impact on development. The ECLAC notes that the average income per capita is only one fifth of what is available now in the United States. Also reported are the lower growth rates for Latin America (1%) compared to those of Europe and the Soviet Union (3–4%). According to the ECLAC, slow development is due partly to *the exhaustion of growth towards the interior of the country*, sparked by the lack of reforms that would have expanded the domestic market (such as land reform, which would have broadened the market for classes not included fully in the economy).

In "strategic" sectors such as autos or steel, transnational corporations were welcomed as providers of needed technology and capital within the import substitution industrialization model. In Table 4.1 above, we can see the significant role played by the multinational corporations in manufacturing. In or about 1970, 24% of manufacturing in Argentina, 50% in Brazil, 30% in Chile, 43% in Colombia, 35% in Mexico, 44% in Peru, and 14% in Venezuela was under foreign control (Jenkins 1984).

ISI in Asia

ISI in South Korea

At the launch of its strategy of industrialization, South Korea had no productive sector (intermediate goods, capital goods, machinery). It had no national products or raw materials for export. It was therefore necessary to import all capital goods needed for industrialization until the installation of a relatively autonomous national production system. The financing of these imports was by recourse to external borrowing (Lanzarotti 1986).

It is important to note already at this time that South Korea denied the use of the FDI and wanted to avoid "foreign domination of the national production." The external debt had exploded. To cope with this situation, South Korea chose exports as an engine of growth and development. Imports were tightly controlled and had focused primarily on capital goods and intermediate goods.

Currency management was under the state monopoly and was centralized. The South Korean financial authorities combined overvaluation of the currency (to reduce import costs for businesses) and subsidies to exporters who benefited from dumping and allowed them to offset losses (they sold off at prices that did not even cover their production costs) (Park 2001).

For the same product, the domestic price was much higher than the export price. Consumption was repressed while savings were favored. The resources were channeled to a centralized public banking system.

South Korea had five-year plans during the period of industrialization (1962–1987). It began with light industries (the 1960s) to substitute domestic production for imports but also for export products. During the 1970s, it launched intermediate goods and chemical industries. During the 1980s, it was time for the launching of industrial equipment (machinery) parts and components. *Here we find the ISI pattern, but combined with an exporter model* (Park 2001).

South Korea, by its particular geo-strategic position, has benefited greatly from the Cold War. A triangular alliance between Japan–South Korea–USA was created and South Korea received a huge economic boost from both allies. The U.S. has helped Korea a lot financially during the early stages of its industrialization (the 1950s and early 1960s). The U.S. markets were opened to South Korean products. South Korea has benefited from other "emergency credit" (debt crisis) and from the U.S. support in international financial markets also. For its part, Japan has provided credit, under very favorable terms, capital goods, and technology needed for industrialization (Lanzarotti 1986).

The ISI in Taiwan

Taiwan, with the population size of Chile in the 1960s, licensed a few auto assemblers, including nationally owned firms. These assemblers never did too well, but they were encouraged to substitute certain parts with imports. These large-scale parts reached a high level of excellence and were soon exported. Taiwan became a major exporter of automobile parts and components worldwide (Llu 1969).

Taiwan's heavy industries were targeted as early as 1961–1964, during the Third Plan: Heavy industry holds the key to industrialization as it produces capital goods. At the same time, exportable items such as watches and other electronic products were promoted. During this period, most of the exports came from the import substitution industries. Protection from foreign competition was not lifted. Getting subsidies for exports was extra.

In Taiwan's electronics industry, there is no clear-cut distinction between an import substitution phase and an export promotion phase. Even though the export of electronics products has sped up since the early 1970s, the domestic market for electronics products was still heavily protected through high import tariffs. Whether protection was necessary for the development of local electronics firms is controversial. However, we do observe that the protection of consumer electronics products forced Japanese electronics firms to set up joint ventures with local entrepreneurs and to transfer technologies to local people, which helped to expand their exporting capabilities (Balassa 1971).

After most heavy industries were, in fact, developed (steel, shipbuilding, petrochemicals, machinery), and the second energy crisis occurred (1979), the goals changed.

In 1982, the Taiwan government began to promote "strategic industries" (machinery, automobile parts, electrical machinery, information and electronics) based on six criteria: large linkage effects; high market potential; high technology intensity; high value-added; low energy intensity; and low pollution.

Taiwan, with the world's second highest growth rate of exports, also tied subsidies to exporting. Cotton textiles, steel products, pulp and paper, rubber products, cement, and woolen textile industries all formed industry associations and agreements to restrict domestic competition and subsidize exports. Permission to sell in Taiwan's highly protected domestic market was made conditional on a certain share of production being sold overseas (Llu 1969).

Lessons Learned from Asian ISI

The inward-looking variant of ISI wasn't totally adopted by most nations in East Asia. Most East Asian countries, while rejecting the inward-looking component of classical import substitution policies, also maintained high tariff barriers. The strategy followed by those countries was to focus subsidies and investment on industries which would make goods for export, and not to attempt to undervalue the local currency. However, outward-looking development was supported by the US incentives for industrialization, as a means of creating a "contention belt" of capitalist countries around the communist nations in Asia.

Conclusions from the Above Analysis

Five stages for the application of an ISI process can be deduced as follows.

First Stage

The production of simple consumer goods should be reduced. For this, the country uses resources and funds available to develop businesses requiring low technical skills, such as textile industries.

Second Stage

This is the stage of mechanization where imports are always being reduced, including some equipment, like, e.g., the production of looms. The industry goes up the ladder by producing higher quality goods.

Third Stage

The role of the state increases: it requires the diversification of industries, including the creation of heavy industries such as cement and steel, which are costly but profitable over several decades. The country should start exporting when the domestic market doesn't absorb the supply.

Fourth Stage

The country produces durable consumer goods such as appliances, consumer electronics, and automotive, then attempts to increase the share of the weight of the middle class consumer of these products, and open the economy to foreigners for encouraging the inflow of multinational firms.

Fifth Stage

Diversification continues until the country is able to export goods produced by the high-tech industry, which requires heavy investments.

Renewal Analysis

In recent years, the strong growth of China has led to a rise of raw material prices accompanied by a decline in the prices of manufactured goods. Therefore, analysis based on the deterioration of terms of trade loses its relevance today for many developing nations with

- The rapid industrialization of the emerging countries (with the notable exception of Africa).
- Export of manufactured goods from the emerging countries.
- The development of the service economy, particularly in the northern countries.
- Deterioration of the trade balance of some northern countries.
- Increasing demand for biofuels increases the need for raw materials such as sugar, oilseeds, and cereals.
- The financial crisis in 2008 positioning food products as refuge for some investments.

Increasingly the so-called least developed countries provide mixed exports. A part of Africa and few countries in other continents, however, can still be considered relevant (as indicated in the FAO report). From this point of view of geo-economy, strong transition seems well set against the phenomenon of declining terms of trade.

Government Incentives

The lower the level of a country's economic development, greater is the need for government intervention to regulate the economy in order to achieve its overall development goals. This involvement can be through setting up an incentives system for supporting companies. The incentives system is, in turn, designed to achieve the country's wider development goals (UN 1992). In this paper ISI policy is seen from the point of view of the company in the host country. Therefore the goals and the incentives system are already designed and the company should get the maximum benefit from the available systems. The incentives system is administrated by some organization or a group of organizations. We focus only on the incentives from which the company can directly benefit. Other incentives of technology transfer, like promoting in-house R&D capacity, developing human capital, including engineering and management skills, protecting the intellectual property, and infrastructure development, act on the whole system including the company in question. In this paper, we are more interested in looking into ISI oriented support tools which are protectionist in nature.

The government provides support to help businesses in order to:

- Start up, grow, and succeed
- Overcome current financial and economic challenges
- Innovate
- Trade internationally

There is a wide range of government support available to businesses, not only through grants and other funding, but also through numerous advisory, guidance, information and other services, including training.

The kinds of help governments offer companies to invest in industry include:

- Grants, which are the money given to the company as a cash item, may be offered for activities such as Training, Employment, Export Development, Recruitment, or Capital Investment projects.
- Loans are funds given to the company which have to be paid back over time.
- Infrastructure improvements provided e.g., roads, electricity supply, telecommunications, and water supply.
- Advice and Information to those searching the essential information required to develop products, services, and markets, since it can be time-consuming and costly for smaller enterprises to do it themselves. Therefore, the government institutions support the companies by providing the needed information.
- Packaged Assistance provided as specific support for different projects and different industries.

Among the ISI instruments, the above listed constitute the industrial incentives as described.

The general orientation of the incentive device could be described in the following formula:

- More the investment is of interest to the national economy, more will be granted and in more significant amounts.
- These benefits vary depending on the location and nature of the investment (e.g., the less developed regions require more for the developing of infrastructure in support of their industrialization efforts).

The Government Incentives System of Algeria

The company selected for an application of the developed framework is in Algeria, therefore the government incentives system in Algeria is described with respect to the ISI instruments and the above organization.

In Algeria, the government incentives are organized into three major schemes:

- The general scheme for the current investment projects located outside those areas to be developed.
- The scheme of development areas concerns the current investment projects located in the said areas.
- The scheme of the investment agreement for the investment projects of particular interest to the national economy.

Benefits Granted Under the General Scheme

Setting-Up Phase of the Project

- Exemption of Value Added Tax (VAT) on goods and services not included.
- Exemption of customs duties on imported equipment not included.
- Exemption of transfer duty on property purchases.

Operational Phase of the Project

- Exemption for 3 years from the Tax on Corporate Profits (TCP)
- Exemption for 3 years from the Tax on Professional Activity (TPA)

Benefits Granted Under the Scheme of Development Areas

Setting-Up Phase of the Project

- Exemption of VAT on goods and services.
- Exemption of customs duties on imported equipment.
- Exemption of transfer duty on property purchases.

- Registration fee at reduced rate (0/00) for the constitutive acts and capital increases.
- Possibility of access to partial or total support for expenditures related to infrastructure works necessary to achieve the investment.

Operational Phase of the Project

- Exemption for 10 years of the TCP.
- Exemption for 10 years of the TPA.
- Exemption for 10 years of the Real Property Tax (RPT).
- Possibility to grant other benefits (carry over losses and depreciation periods).

Benefits Granted Under the Scheme of the Convention

Investments under the scheme can benefit from all or part of the following benefits:

Setting-Up Phase of the Project (Five Years Maximum)

- Exemption of duties, taxes, charges, and other levies on all goods and services imported or purchased locally.
- Exemption of transfer duty on property acquisition and legal advertisements.
- Exemption of registration fees.
- Exemption of the RPT.

Operational Phase (Ten Years Maximum)

- Exemption of the TCP.
- Exemption of the TPA.
- Besides these advantages, the National Investment Council (NIC) may provide other facilities or benefits for investments in sectors of particular interest to the national economy.

Framework Development

In this paper, it is assumed that the perspective of business research can be a very useful complement to economics. If one attempts to combine the insights of economic and business research perspectives, one quickly becomes aware of the differences between them. Economic policy research, and in particular trade and macroeconomic research, often tends to be directed at issues and challenges that arise at the level of regions, nations, or even groupings of nations such as the EU. Business policy research, on the other hand, tends to be focused on the performance of individual firms or groups of firms. This distinction was highlighted by Porter in his analysis of the competitive advantage of nations, when he stressed the point that it is more helpful to consider firms as competing in industries, not in nations (Porter 1990). Kotler et al. attempt to connect macroeconomic policy with microeconomic behavior of industries, firms, and consumers, as well as to apply strategic planning to the building of national wealth (Kotler et al. 1997). The interest is to find out how the firms behave when subjected to changes in the wider external policy environment. Business research frameworks, on the other hand, tend to be concerned with the analysis of the consequences of management actions directed at improving the prospects of a "specific" firm within a given (usually fixed) external policy environment.

Because of this very basic difference in the main emphasis of their disciplines, economic and business researchers often tend to misunderstand, discount, or ignore each other, and sometimes adopt quite dismissive attitudes toward each others' methodologies.

At the level of the individual firm or corporation, a strategy is usually formulated in a context where government policies are largely exogenous, and firms address the challenges of assessing the business portfolio, identifying strategic goals, and redefining the business domain.

The crucial role of management is to formulate a corporate strategy that aligns with the nation's wealth-building strategy (Kotler et al. 1997).

In this paper, we opt for ISI as a macropolicy and, therefore, look at how a firm should operate within such environment to avoid any conflict between the macro- and micropolicies. Many pieces in the literature cover the ISI strategy at the national or regional level, but none of them considers the building-up of the company strategy respecting the fact that the considered firm has its business in the country opting for the ISI strategy.

How do business researchers delineate their field of enquiry? At the beginning of most textbooks on strategic marketing, (Finch 2004) there is usually an explanatory diagram made up of three concentric circles, labeled broadly as follows:

The outer business environment; It contains all the economic and competitive forces that make up the domain of the economic policy researcher, in addition to the forces of technical progress, social forces, legal issues, environmental protection rules, and political forces (such as ideology).

The middle circle of business strategy is the domain where much business-oriented research is focused. Key issues include the determination of a company's competitive advantage within a given outer business environment. *It also includes the selection of target markets and the positioning of products and brands within the selected markets.*

The inner circle is the environment of business tactics and it encompasses all aspects of the so-called "marketing mix." Here one has moved away from the wider role of public "environmental" policy and medium to long-term business strategy, and the emphasis is placed on the immediate actions of managers within individual firms (e.g., pricing policy, product development, promotion activities, and distribution channels).

In the framework being developed, two concentric circles are considered:

- The outer business environment through the national industrialization strategy, which is the ISI in our case, is characterized by the government incentives system and the protectionism approach.
- The middle circle is represented by: customs data (another form of the market need), the state of the domestic supply chain (DSC), and by the general status of the considered company.
- However, the inner circle is clearly described in the implementation of the developed strategy based on the outer and middle circles.

First: SWOT Analysis of the company is made in formulating the firm's *strategic vision*. This involves inserting the company into its appropriate competitive position in the global/national marketplace, applying a SWOT-type analysis at a microlevel to access its capabilities, and settling on its strategic thrust, or how it can achieve its stated goals.

Second: the range of developing appropriate *strategic postures* through a business strategy view of a range of policy initiatives that are mainly economic in nature, e.g., approaches to investment (including R&D, equipment, training, and development of new products); spatial policies to move to another region, where there is more aid and less competition, trade policies: new marketing strategy, increase the production (In this paper only new manufacturing products are detailed).

Third: aspects of strategic implementation, which require consideration of the abilities of company to execute strategic policy and to function in coherence with national policy (even though it was already considered in the setting up of the strategy). There is a need to be aware of the overall atmosphere to ensure cooperation in the national arena.

The developed framework is a sequence of three main stages as described in Fig. 4.1:

- First stage: elaboration of the state of affairs
- Second stage: utilization of the gathered data as the first step in making decisions and choices using very formal algorithms
- Third stage: execution of the selected solutions

First Stage: Elaboration of State of Affairs

Supply Segmentation

A product-market is segmented, that is, it is cut into subset of products sharing characteristics with the currently manufactured products by the company interested in developing new products.

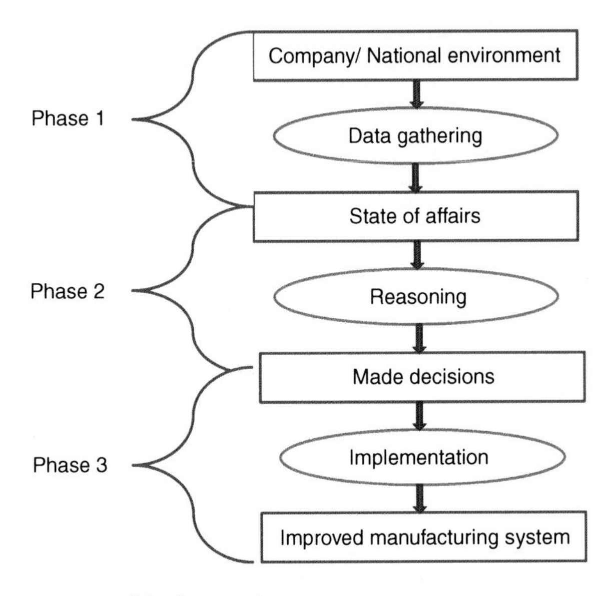

Fig. 4.1 Sequence stages of the framework

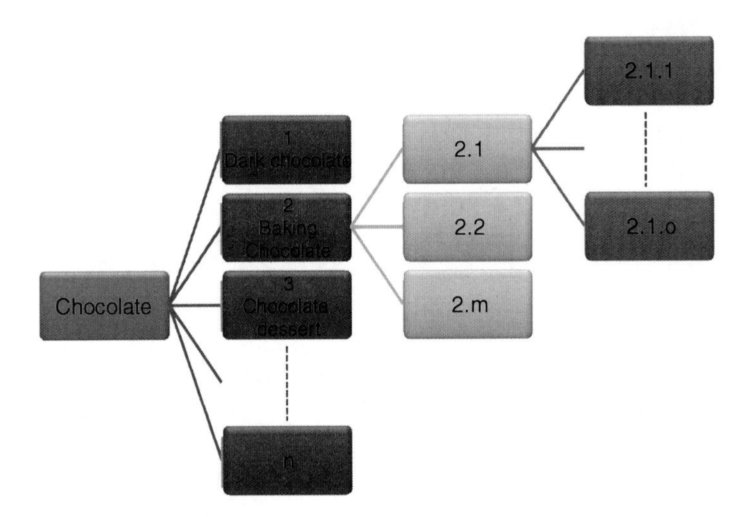

Fig. 4.2 Chocolate supply segmentation

Figure 4.2 shows a simplified example of a supply segmentation of chocolate.

We note in this example the basic product "chocolate" could have different types targeting different needs of customers. The considered company manufactures one or more of these products. By the supply segmentation we answer the question "what are the most eventual products to be manufactured by the company in question?" Because we assume that the products of the same segmentation are very similar, they do not require much new knowledge and, therefore, the absorptive capacity is high. This would make the project successful.

Product n.m.o			
Tariff x% VAT y%			
Imports CIF		Exports FOB	
Value in K Euros year	Mass in Tons year	Value in K Euros year	Mass in Tons year
Average between year- year+10	Average between year- year+10	Average between year- year+10	Average between year- year+10
Cost index (Value /Mass)		Cost index (Value /Mass)	

Fig. 4.3 International trade module of the product n.m.o

International Trade Module Assignment

The foreign trade statistics compiled by customs include imports and exports by regions and countries and types of products. Through these statistics, it is possible to have access to quantities of goods imported or exported, their origins or destinations, and their values. The tariff is notified by the customs nomenclature.

For each product of the segmentation tree, an international trade module is assigned and it has the following form (Fig. 4.3):

Cost index allows comparing the price of the imported product with the exported one; when there is no export, it is then compared with similar products manufactured by the domestic company. Cost index is calculated with the taxes (tariffs and VAT).

It is important to have the statistics of exterior trade of a relatively long period because 1 year is not enough to know if the imported product is needed periodically to avoid investing for a not really needed product, because in this case the local manufacturing of the product will not be profitable especially if the investment is expensive.

Drawing of Pareto diagram for all the imported products determines the ABC-classification which groups a range of imported products into three categories (A,B,C) where each category should be handled in a different way.

- References products representing up to 80% of the imports values: category A
- References products representing 80–95% of imports values: category B
- References products representing 95–100% of imports values: category C

SWOT Matrix Filling

SWOT is an acronym for *Strengths*, *Weaknesses*, *Opportunities*, and *Threats*. These four factors make up the SWOT Matrix (Finch 2004).

A SWOT analysis will be used to measure the company's *competencies* and identify *opportunities* to be taken from the government incentives.

SWOT	Intern	Extern
Positive	Strengths	Opportunities
	R_f	Govi, Ne, DSC, Cp
Negative	Weaknesses	Threats
	R_f	DCS

Fig. 4.4 Adapted SWOT matrix

Strengths: attributes of the company which are helpful in achieving the manufacturing of one or more of the imported products. They are described in the vector R_f (Recipient firm or the transferee firm) encompassing the status of the considered company. Their coordinates are relevant for the technology Transfer projects. When the coordinates of the vector R_f are negative it represents a weakness.

Weaknesses: attributes of the company which are not helpful.

Opportunities: helping attributes which are exterior to the company. For the substituting project, we consider the developed DSC, the government incentives (Govi) of the ISI approach, the low competition (Cp) with imported products because of the incentives and the market need (Ne) as opportunities.

Threats: external attributes threatening the success of the project. A less developed DSC could represent a real threat for local manufacturing (Fig. 4.4).

R_f (L, F, P, Te, Exp) is the vector describing the status of the company interested in locally manufacturing currently imported products where:

- L: company labor characteristics; price (cheap or expensive), its technical level, and technical ability to absorb new knowledge.
- F: financial situation of R_f; its ability to invest in the project and for which amortization period (long, short).
- P (p_1, p_2, p_3, p_4, p_5): describes the current production system where p_1 is Plant strategy, p_2 process choice, p_3 production system choice, p_4 operational methods choice, and p_5 is human resource policy choice.
- Te: existing technology level and the absorption capacity of new technology. Absorption capacity is considered to be high because the products of the same segmentation are similar and thus they do not really require new knowledge.
- Exp: experiences with ITT projects (number, success, or failure).

Govi (Ts, Tx) is the vector describing the government incentives. In ISI strategy the government incentives are designed to replace imported products with domestic production.

- Ts: aids related to the phase of setting up the project
- Tx: aids related to the exploitation phase

Ne (Qte, per) is the vector representing the market need of the imported product, it is considered in our case always as an opportunity for the ITT project, because we

Table 4.2 BOM matrix

	Is composed of	Product references	
Supply		1.1	n.m.o
	a		5
	b		
	c	2	

consider that the locally manufactured product will be similar and eventually cheaper (because of the incentives), thus the distribution of the product will be easy. In other words the market need for the imported product is converted into a need of the same product locally manufactured. Qt is the needed quantity for a period per se. Ne is directly inferred from the international trade modules.

DSC (Av, C, Q); in this paper it is comprised only of the R_f suppliers, because the customers are considered in the market need. The DSC is concerned with the ability of the domestic suppliers to supply the required raw materials, parts, and other items necessary for the manufacturing of the currently imported products. DSC is enough when the suppliers are able to supply at least most of the necessary items at the required time for the required quantities and quality. So when the DSC is not developed enough, it could represent a threat for local manufacturing, otherwise it is considered an opportunity.

- Av: the required matters for production (raw materials, parts, energy, services...) are locally available or imported but without increasing the cost of the product.
- C: the cost of the local supplies.
- Q: the quality of the local supplies.

Cp: is an indicator of the eventual competition with the imported product. It is supposed to be low (in an ISI approach) because of imposed taxes on the imports. The locally manufactured product is assumed to have at least the same quality of the imported one, otherwise the Cp would be high, and in this case it represents a threat.

Bill of Material (BOM) for the Product of the Supply Segmentation

BOM is the list of all raw materials, parts, intermediates, sub-assemblies, etc., (with their quantities and description) required to construct, overhaul, or repair the product. BOM will allow for the determination of the required supplies for the manufacturing of the currently imported products and thus to evaluate the domestic suppliers of the required supplies.

Because the products are from the same supply segmentation, they have many common constituents. Therefore, the BOM matrix is suggested (Table 4.2).

Assignment, when it is possible, of the international trade module for each object (constituting the imported product) is done in order to know at least if the object is available locally (when it is not imported or when it is exported).

A questionnaire to be answered by the domestic suppliers is prepared in order to inform the coordinates of the vector DSC, such as the quantity, quality, and costs of the domestic supplies.

In Table 4.3, most of the important criteria related to the capacity, quality, and the cost of supplied items are listed:

Table 4.3 helps in building the content of the questionnaire.

Second Stage: Utilization of the Gathered Data to Make Decisions

After gathering the required data and writing them on the form of international trade modules, SWOT matrix, and MOB matrices, these data are used to make strategic choices. The reasoning process of data utilization is described in the following organization chart (Fig. 4.5).

1. If the current manufactured product is still imported:
   - Answer the questions concerning its quality and quantity: less quality or less quantity?
     - Increase quality and/or quantity by investing in new technologies or just using more efficiently the current manufacturing system
     - Or reviewing the marketing policy
   - When the product is exported then the firm should develop its capacities to start exporting
2. When the currently manufactured product is not imported (not mentioned in the statistics) or only for very few quantities for not reasonable reasons (some customers prefer the imported products)
   - Check the other similar products:
   - When they aren't imported, are they:
     - Locally manufactured by other manufacturers
     - Or there is no need, therefore there is a possibility to create a need
   - When are they imported:
   - For which quantity and cost: set the Pareto graph
   - R_f, is it able to manufacture them locally:
     - See the SWOT matrix:
       - Financial situation
       - Technology absorption (similar products)
       - Government aids: protectionism system

Table 4.3 Supply chain criteria

Criterion	Reference(s)	Remarks
Physical criteria		
Delays (DL)	P. Finch, "Supply chain risk management," Supply Chain Management: An International Journal, 2004, Vol. 9, pp. 183–196	Due to high utilization or another cause of inflexibility of suppliers
Disruptions (DS)	P. Kotler, S. Jatusripitak and S. Maesincee. "The Marketing of Nations," New York, The Free Press, 1997; P. Finch, "Supply chain risk management," Supply Chain Management: An International Journal, 2004, Vol. 9, pp. 183–196; J. Hallikas, I. Karvonen, U. Pulkkinen, V. Virolainen, M. Tuominen et al. "Risk management processes in supplier networks," International Journal of Production Economics, 2004, Vol. 90, 2004, pp. 47–58	Very unpredictable but of high impact
Supplier capacity constraints (CC)	H. Peck, "Drivers of supply chain vulnerability: an integrated framework," International Journal of Physical Distribution & Logistics Management, Vol. 35, 2005, pp. 210–232; J. Martin, "Blocage de développement et industrialisation par substitutions d'importations. L'exemple de l'Argentine," Tiers Monde, 1967, Vol. 8, No. 30, pp. 503–515	Unable to handle sudden spurt or to utilize excess slack
Production technological changes (TC)	G. Zsidisin, Panelli A, Upton R et al., "Purchasing organization involvement in risk assessments, contingency plans, and risk management: an exploratory study," Supply Chain Management: An International Journal, 2000, Vol. 5, pp. 187–198.; L. Giunipero, R. Eltantawy, "Securing the upstream supply chain: a risk management approach," International Journal of Physical Distribution & Logistics Management, 2000, Vol. 34, 2000, pp. 698–713	Supplier not able to produce items to necessary demand level and at a competitive price
Transportation (TR)	L. Giunipero, R. Eltantawy, "Securing the upstream supply chain: a risk management approach," International Journal of Physical Distribution & Logistics Management, 2004, Vol. 34, pp. 698–713.; G. Zsidisin, Panelli A, Upton R et al., "Purchasing organization involvement in risk assessments, contingency plans, and risk management: an exploratory study," Supply Chain Management: An International Journal, 2000, Vol. 5, pp. 187–198	Pertinent in case of logistics outsourcing as is the case of 3PL

(continued)

Table 4.3 (continued)

Criterion	Reference(s)	Remarks
Physical criteria		
Inventory (IN)	H. Peck, "Drivers of supply chain vulnerability: an integrated framework," International Journal of Physical Distribution & Logistics Management, 2005, Vol. 35, pp. 210–232; R. Speckman, E. Davis, "Risky Business: expanding the discussion on risk and the extended enterprise." International Journal of Physical Distribution & Logistics management, 2004, Vol. 34, pp. 414–433; G. Svensson, "Key areas, causes and contingency planning of corporate vulnerability in supply chains: A qualitative approach," International Journal of Physical Distribution & Logistics Management, 2004, Vol. 34, pp. 728–748.; H. Peck, "Drivers of supply chain vulnerability: an integrated framework," International Journal of Physical Distribution & Logistics Management, 2005, Vol. 35, pp. 210–232	Excess inventory for products with high value or short life cycles can get expensive
Procurement (PR)	H. Peck, "Drivers of supply chain vulnerability: an integrated framework," International Journal of Physical Distribution & Logistics Management, 2005, Vol. 35, pp. 210–232	Unanticipated increases in acquisition costs
Capacity Inflexibility (CI)	J. Hallikas, I. Karvonen, U. Pulkkinen, V. Virolainen, M. Tuominen et al. "Risk management processes in supplier networks," international Journal of Production Economics, 2004, Vol. 90, pp. 47–58	Facility fails to respond to changes in demand
Design (DG)	P. Finch, "Supply chain risk management," Supply Chain Management: An International Journal, 2004, Vol. 9, pp. 183–196	Suppliers' inability to incorporate design changes
Poor quality (PQ)	G. Svensson, "Key areas, causes and contingency planning of corporate vulnerability in supply chains: A qualitative approach," International Journal of Physical Distribution & Logistics Management, 2004, Vol. 34, pp. 728–748.; G. Zsidisin, Panelli A, Upton R et al., "Purchasing organization involvement in risk assessments, contingency plans, and risk management: an exploratory study," Supply Chain Management: An International Journal, 2000, Vol. 5, pp. 187–198. G. Zsidisin, Panelli A, Upton R et al., "Purchasing organization involvement in risk assessments, contingency plans, and risk management: an exploratory study," Supply	If suppliers plants don't have quality focus

	Chain Management: An International Journal, 2000, Vol. 5, pp. 187–198.; M. Treleven, S. Schweikhart, "A risk/benefit analysis of sourcing strategies: Single vs. multiple sourcing", Journal of Operations Management, 1988, Vol. 7, pp. 93–114	
Financial criteria		
Cost/price (CR)	M. Treleven, S. Schweikhart, "A risk/benefit analysis of sourcing strategies: Single vs. multiple sourcing," Journal of Operations Management, 1988, Vol. 7, pp. 93–114.; G. Zsidisin, Panelli A, Upton R et al., "Purchasing organization involvement in risk assessments, contingency plans, and risk management: an exploratory study," Supply Chain Management: An International Journal, 2000, Vol. 5, pp. 187–198	Concerns competitive cost risk
Untimely payments (UT)	H. Peck, "Drivers of supply chain vulnerability: an integrated framework," International Journal of Physical Distribution & Logistics Management, 2005, Vol. 35, pp. 210–232.; H. Peck, "Drivers of supply chain vulnerability: an integrated framework," International Journal of Physical Distribution & Logistics Management, 2005, Vol. 35, pp. 210–232	Loss of goodwill and may impact much on SMEs
Settlement process disruption (SP)	J. Cavinato, "Supply chain logistics risks," International Journal of Physical Distribution & Logistics Management, 2004, Vol. 34, pp. 383–387	Leads to delay in payments and impacts SC profitability
Lack of hedging (LH)	H. Peck, "Drivers of supply chain vulnerability: an integrated framework," International Journal of Physical Distribution & Logistics Management, 2005, Vol. 35, pp. 210–232	Disastrous in case of bankruptcy of partners in the SC
Unstable pricing (UP)	H. Peck, "Drivers of supply chain vulnerability: an integrated framework," International Journal of Physical Distribution & Logistics Management, 2005, Vol. 35, pp. 210–232	May lead to lack of trust among SC partners

Source: Wu (2009)

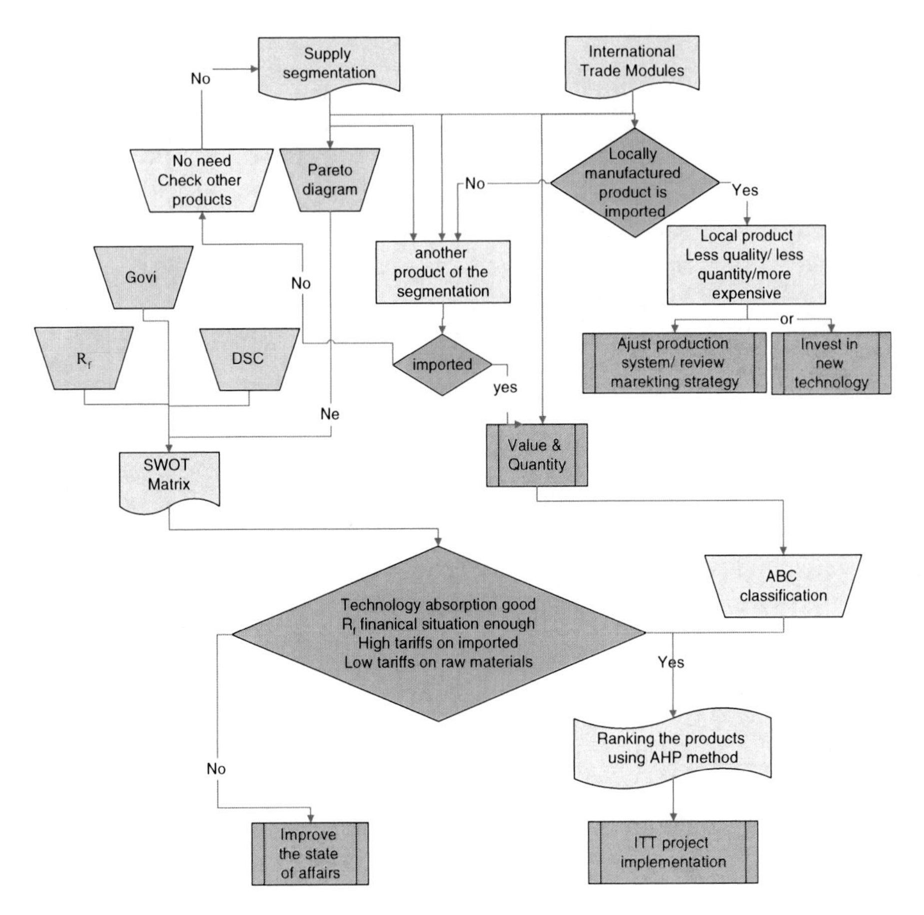

Fig. 4.5 Algorithm chart of the decision-making process

  - High tariffs on the imported products, low tariffs on the raw materials, parts and equipment
- DSC/statistics/government incentives (protectionist system tariffs mostly):
  - Make the BOM of the future manufactured products: to know about their components and raw materials and then check the local supplier abilities by filling the questionnaire
    - When not enough developed DSC – Start by assembling parts and then the next steps manufacturing the parts or develop local suppliers
  - Assign for each possible product the corresponding government aids (maybe different tariffs on the imported product)
- Rank the products likely to be manufactured using AHP methodology and the criteria are: Pareto graph, investment costs, risks: complexity of the technology, required time…

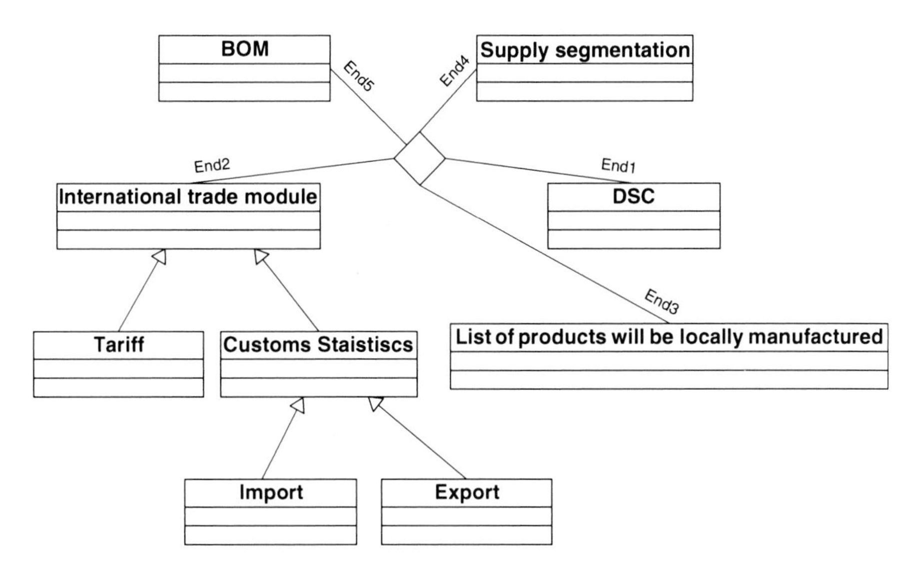

Fig. 4.6 Class diagram for the implementation of the framework

There is no real need for implementing the algorithm when it is used at the company level. However, the same framework could be extended at an upper level for eventual utilization by the national government or regional one to foster the promising industrial sectors. In this case the framework should be implemented because of the huge amount of data. In Fig. 4.6, a rough UML class-diagram of the framework programming can be seen.

Third Stage: Implementation of the Selected Choices

In this stage, the results of the last stage are applied. In other words, one performs an international transfer project (ITT) to acquire the necessary technology allowing local manufacturing of the selected products.

The situation of the company and its environment are already reviewed and therefore this information will be used to conceptualize the future manufacturing system, find a technology supplier, develop the human resources, implement the new manufacturing system in the company, and adapt it for the local conditions.

The execution of an ITT project is already developed by Bessam et al. (Bessam et al. 2010). It was modeled using IDEF0 approach (Fig. 4.7).

The above ITT project implementation is modeled using IDEF0 methodology. It is a static description focusing on the inputs which are transformed into outputs using relevant methods, procedures, and resources. Each box of the diagram represents a function of the process described by its inputs, outputs, controls, and resources.

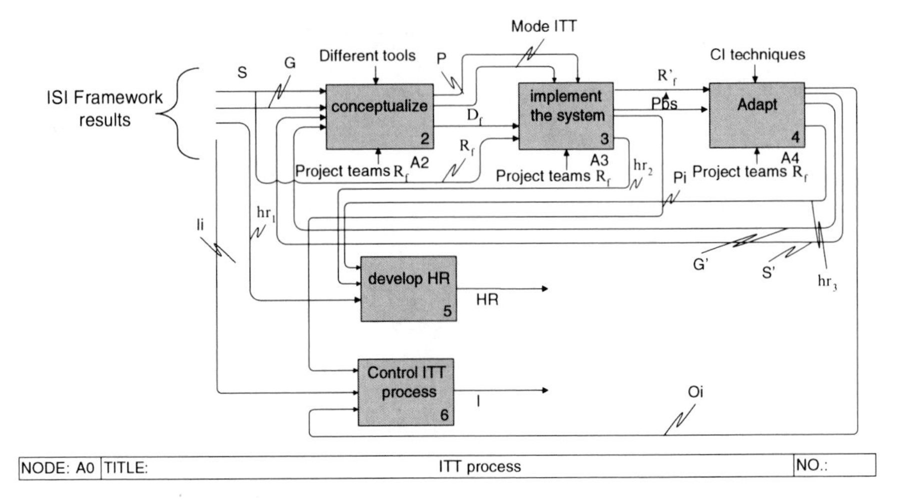

Fig. 4.7 ITT project implementation

The advantage of such a model is the fact that it is the first model gathering most of the ITT project objects in the form of vectors evolving during the project stages. Details of the implementation stage are presented in Bessam et al. (2010).

Case Study: The Cannery SME in Algeria

The Algerian government has adopted an ISI policy after its independence to stimulate the industrialization of the country (Schiephake 1975).

Application example: Algeria (late 60s–early 80s)

- Socialist and nationalist Plan
- Petroleum exporting country
- Focus on steel, petrochemicals, and infrastructure
  - Results:
    Industrial production increased threefolds between 1970 and 1985
    Sonatrach became the largest company in Africa
  - Problems:
    High oil revenues needed for investment
    Projects often oversized
    Limited job creation
    Labor not trained for new technologies: lack of engineers and managers
    Consumer industries and agriculture are neglected
    Algeria became an importer of food

After 2000, Algeria adopted once again an industrial strategy, where the ISI-oriented policy chosen was based on the protectionism and fostering of equipment imports. Foreign direct investment (FDI) was highly restricted by the law 51–49. Public investments were mostly in developing infrastructures. In 2006, the industrial equipment imports represented 39% with food imports taking up 17% and consumer goods 14%.

Although Algeria is a petroleum exporter (oil and natural gas revenues in 2007 being more than $70 billion according to the Algerian national statistics institute), the ISI approach is still suitable for application. As long as the general trend of oil prices was increasing, the phenomenon of trade terms deterioration wouldn't arise. Yet, in many oil exporter countries like Algeria, the production was mostly steady or decreasing, while the imports of manufactured goods were increasing steadily. For Algeria, oil and gas constitute 96% of the country's exports.

We take a simple example from the food industry, which is relatively old. We take an SME company because it represents the case of the most private sector business and, therefore, the industrial strategy of the country is concentric on the SMEs.

The case study is about the company Conserverie Royal (CR), which is a cannery, specialized in canned meat and vegetables. It is located in a very active region around Algiers and is interested in developing and expanding its product range, especially in tomato products because the raw material is cheap and available in the region.

In the following, a simplified explanation is provided of the framework developed to select what the eventual products to be in the future manufactured by CR should be, knowing that the national strategy is ISI-oriented.

First Stage

Supply Segmentation

The aim is to build the product supply segmentation for the industrial preparation of tomatoes in Algeria. The segmentation could be more developed, but we do not have more precise data in the customs nomenclature statistics (Fig. 4.8).

The currently manufactured products by the CR Company are 1.1 and 1.2. The percentages in the third layer of the chart represent tomato content in the product.

International Trade Modules Assignment

In the following, all the modules of international trade corresponding to the products of the supply segmentation are presented (Table 4.4–4.11):

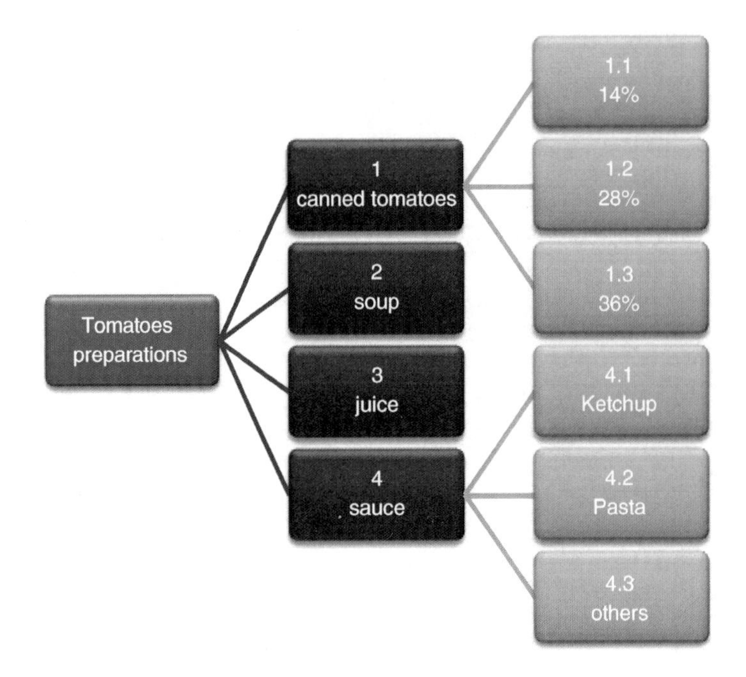

Fig. 4.8 Supply product segmentation

Table 4.4 International trade module of product 1.1

Tariff 30% VAT 17%			
Imports CIF		Exports FOB	
Value in K Euros 2008	Mass in Tons 2008	Value in K Euros 2008	Mass in Tons 2008
0		0	
Average between 2000 and 2008	Average between 2000 and 2008	Average between 2000 and 2008	Average between 2000 and 2008
0		22	31
Cost index (value/mass)		Cost index (value/mass)	
		0.71	

Table 4.5 International trade module of product 1.2

Product 1.2			
Tariff 30% VAT 17%			
Imports CIF		Exports FOB	
Value in K Euros 2008	Mass in Tons 2008	Value in K Euros 2008	Mass in Tons 2008
0		320	301
Average between 2000 and 2008	Average between 2000 and 2008	Average between 2000 and 2008	Average between 2000 and 2008
0		290	279
Cost index (value/mass)		Cost index (value/mass)	
		0.71	

Table 4.6 International trade module of product 1.3

Product 1.3			
Tariff 30% VAT 17%			
Imports CIF		Exports FOB	
Value in K Euros 2008	Mass in Tons 2008	Value in K Euros 2008	Mass in Tons 2008
61	59	0	
Average between 2000 and 2008	Average between 2000 and 2008	Average between 2000 and 2008	Average between 2000 and 2008
66	64	0	
Cost index (value/mass)		Cost index (value/mass)	
1.031			

Table 4.7 International trade module of product 2

Product 2			
Tariff 30% VAT 17%			
Imports CIF		Exports FOB	
Value in K Euros 2008	Mass in Tons 2008	Value in K Euros 2008	Mass in Tons 2008
20	37	0	
Average between 2000 and 2008	Average between 2000 and 2008	Average between 2000 and 2008	Average between 2000 and 2008
35	41	0	
Cost index (value/mass)		Cost index (value/mass)	
0.85			

Table 4.8 International trade module of product 3

Product 3			
Tariff 30% VAT 17%			
Imports CIF		Exports FOB	
Value in K Euros 2008	Mass in Tons 2008	Value in K Euros 2008	Mass in Tons 2008
27	56	0	
Average between 2000 and 2008	Average between 2000 and 2008	Average between 2000 and 2008	Average between 2000 and 2008
21	49	0	
Cost index (value/mass)		Cost index (value/mass)	
0.42			

Table 4.9 International trade module of product 4.1

Product 4.1			
Tariff 30% VAT 17%			
Imports CIF		Exports FOB	
Value in K Euros 2008	Mass in Tons 2008	Value in K Euros 2008	Mass in Tons 2008
420	125	0	
Average between 2000 and 2008	Average between 2000 and 2008	Average between 2000 and 2008	Average between 2000 and 2008
390	111	0	
Cost index (value/mass)		Cost index (value/mass)	
3.51			

Table 4.10 International trade module of product 4.2

Product 4.2			
Tariff 30% VAT 17%			
Imports CIF		Exports FOB	
Value in K Euros 2008	Mass in Tons 2008	Value in K Euros 2008	Mass in Tons 2008
611	504	52	59
Average between 2000 and 2008	Average between 2000 and 2008	Average between 2000 and 2008	Average between 2000 and 2008
643	511	47	41
Cost index (value/mass)		Cost index (value/mass)	
1.25		1.14	

Table 4.11 International trade module of product 4.3

Product 4.3			
Tariff 30% VAT 17%			
Imports CIF		Exports FOB	
Value in K Euros 2008	Mass in Tons 2008	Value in K Euros 2008	Mass in Tons 2008
320	281	11	13
Average between 2000 and 2008	Average between 2000 and 2008	Average between 2000 and 2008	Average between 2000 and 2008
401	389	9	11
Cost index (value/mass)		Cost index (value/mass)	
1.03		0.81	

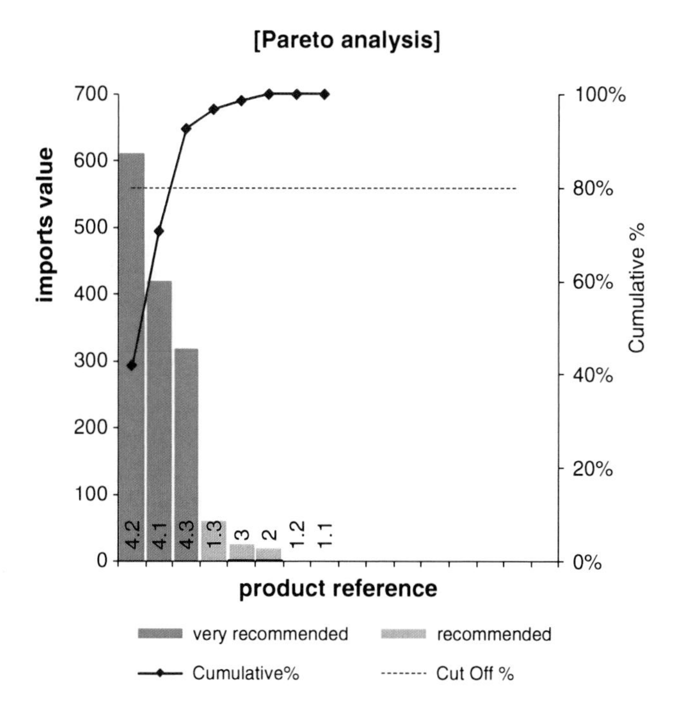

Fig. 4.9 Pareto analysis of the tomato products

The Pareto diagram shows clearly that the products of branch four are the most interesting. The most important products in term of imports value (more than 80%) have the references 4.2, 4.1, 4.3. In this case the conclusion is obvious. However, the advantage is significant in case there are thousands of products, because the selection could be easily computerized (Fig. 4.9).

The SWOT Matrix

R_f (L, F, P, Te, Exp):

- L: cheap, medium, high.
- F: the company CR is able to invest not more than 400 K Euros for 5 years amortization period maximum.
- P: the production system is relatively new, 15 years passed since its installation and it is well maintained.
- Te: medium technology level, high ability of absorption (skilled workforce, long experience).
- Exp: 2 ITT projects already performed.

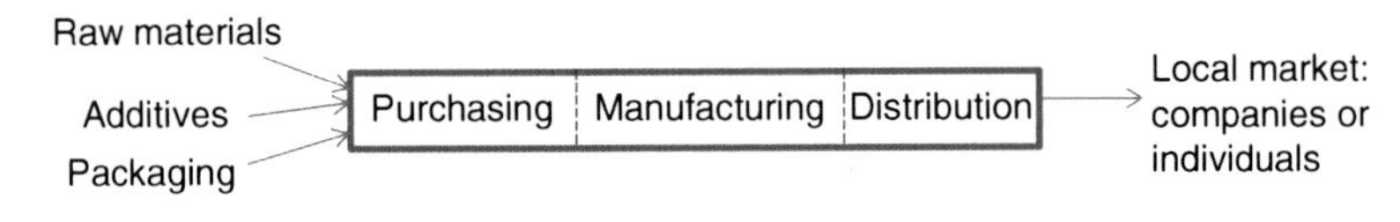

Fig. 4.10 General BOM of the imported products

Govi (Ts,Tx): Government incentives of Algeria are detailed in the § XI.1.

Ts and Tx are highly positive, reflecting the applied ISI policy in the country. However, government incentives cannot serve as criteria of selection, because in this case they are the same for all products.

Ne (Qte, per): Each product has its own Ne, which is directly deduced from the international trade modules.

The products are relatively simple and similar, therefore we opt for the same BOM (Fig. 4.10):

- Raw material: tomatoes/vegetables; available in domestic market in good quality, and enough quantity.
- Packaging: also very well supplied; according to the previous experience of CR. The few imports of packaging confirm.
- Additives: some of them are not locally available but their tariffs are low (5%).

DSC (Av, C, Q): The CR company already has experience with the same raw materials and items necessary for its production, therefore it is able to evaluate the DSC. In this case DSC is not a criterion of selection because it is the same for all products.

The only criterion is the Pareto diagram. Product 1.3 offers the most potential for the company according to the Pareto diagram. Its investment costs are the lowest and the time required the shortest.

Conclusions and Perspectives

The developed framework is based not only on the status of the company but also on its environment. It encompasses all the ITT project aspects from the data for strategy formulation. The main strength of the framework is the fact that it is a syntactic approach, which means it is independent of any semantics.

The developed framework could be computer-implemented for a larger use and treatment of vast amounts of data. The current framework uses quantitative and qualitative information. An advantage for the framework is the fact that it could be used even when there is a lack of quantitative data only by using the assessment of the company based on its experience in the business.

The developed framework is highly connected to the national industrial policy, which increases the harmony between business strategy of the company and the national strategy.

The framework could be improved by considering new parameters and by developing the assessment of the DSC. It does not show the actions to be taken when the current status is not suitable for the local manufacturing of a product, therefore, as a perspective, some modules will have to be developed in order to allow for improving the situation and making it suitable for an ITT project. The use of government incentives is not completely automated, and their modeling should be improved.

References

Aboites J (1987) Industrialisation et développement agricole au mexique: une analyse du régime d'accumulation de long terme (1939-1985). CEPREMAP Working Papers (Couverture Orange) 8727, CEPREMAP

Auroi C (2009) Tentatives d'intégration économique et obstacles politiques en Amérique latine dans la seconde moitié du XXe siècle. Relat Int 137(1):91–113

Bairoch P (1973) Agriculture and the industrial revolution 1700–1914. In: Cipolla C (ed) The industrial revolution – Fontana economic history of Europe, vol 3. Collins/Fontana, London

Balassa B (1971) Industrial policies in Taiwan and Korea. Business Econ 106(1):55–77

Bessam HE, Gadow R, Arnold U (2010) International technology transfer process: model of activities sequence, 4th International Conference on Managing Enterprise of the Future Poznan, Poland

Boix C (2009) Development & democratization. IBEI Working Papers, No 26, Barcelona

Chang H (2002) "Article summarizing the 'Kicking Away the Ladder' book", Post-Autistic Economics Review, UK

Chenery H (1989) Handbook of development economics, vol 2. Elsevier science publisher, Amsterdam

De Pablo J (1977) Beyond import substitution: the case of Argentina. World Develop 5(l):7–17

Destanne de Bernis G (1974) Les industries industrialisantes et les options algériennes. Tiers Monde 12(47):545–563

Fauré L, André Y (2003) Les défis du développement économique et social du Brésil contemporain. Documents de travail, Groupe d'Économie du Développement de l'Université Montesquieu, pp 1–30

Felix D (1965) Monetarists, structuralists, and import-substituting industrialization: a critical appraisal. Business Econ 1(10):137–153

Finch P (2004) Supply chain risk management. Suppl Chain Manage Int J 9:183–196

Gélinas J (1994) Et si le Tiers Monde s'autofinançait – De l'endettement à l'épargne. *Les* Éditions Écosociété, Montréal

Gélinas J (2001) Report of the Commissioner of the Environment and Sustainable Development to the House of Commons. Minister of Public Works and Government Services, Ottawa, Ontario, Canada

Hoshino T (2001) Industrialization and Private Enterprises in Mexico. Institute of Developing Economies, Japan External Trade Organization, Japan

Jenkins R (1984) Transnational Corporations and Industrial Transformation in Latin America, New York, St. Martin's

Karunaratne N (1980) Export oriented industrialization strategies. Intereconomics 15(5):217–223

Kotler P, Jatusripitak S, Maesincee S (1997) The marketing of nations. The Free Press, New York

Lanzarotti M (1986) L'industrialisation en Corée du Sud: une analyse en sections productives. Tiers Monde 27(107):639–657

Llu T (1969) The process of industrialization in Taiwan. Develop Econ 7(1):63–80

Martin J (1967) Blocage de développement et industrialisation par substitutions d'importations. L'exemple de l'Argentine. Tiers Monde 8(30):503–515
Mazumdar D (1991) Import-substituting industrialization and protection of the small-scale: The Indian experience in the textile industry. World Develop 19(9):1197–1213
Meynier A (1953) L'industrialisation du Mexique. Annales de Géographie 62(332):318–320
Nath SK (1962) The theory of balanced growth. Oxford Economic Papers, New Series, vol 14, No. 2, pp 138–153
Nurkse R (1961) Equilibrium and Growth in the World Economy, Cambridge, Mass., Harvard University Press
Park H (2001) Industrialization and Social Mobility in Korea. CDE Working Paper No., pp 99–32
Porter M (1990) The competitive advantage of nations. Macmillan, London
Rollinat R (2005) Analyses du Développement et Théories de la Dépendance en Amérique Latine: L'actualité d'un débat. Cahiers PROLAM/USP 4:97–118
Rosen S, Simon J, Thea D, Zeitz P (1999) Exchanging Debt for Health in Africa: Lessons from Ten Years of Debtfor-Development Swaps. Development Discussion Paper No. 732, Harvard University
Samuel J (1978) Modernization and dependency: alternative perspectives in the study of Latin American underdevelopment. Comparative Polit 10:535–557
Schiephake K (1975) "Regional Development and Oil Strategy: the Case of Algeria", Intereconomics 7(7):202–206
Singer H (1998) "The Terms of Trade Fifty Years Later – Convergence and Divergence", The South Letter 30, 1.
UN (1992) Formulation & implementation of foreign investment policies. Unites Nations Publications, New York
UN (2003) Import substitution industrialization looking inward for the source of economic growth. Puzzle Lat Am Econ Develop 1:51–79
Waterbury J (1999) The long gestation and brief triumph of import-substituting industrialization. World Develop 2(27):323–341
Werner B (1975) Import substitution and industrialization in Latin America: experiences and interpretations. Lat Am Res Rev 7:95–122
Wu T, Blackhurst J (2009) Managing supply chain risk and vulnerability: tools and methods for supply chain decision makers. Springer, New York

Chapter 5
Technology Transfer and the Knowledge Economy

Mahmut Kiper

Introduction

Knowledge is considered as a main driver in today's global economy. The enhanced role of knowledge and its effects on the global economy have changed the definition of technology. While technology was accepted as a commodity in the past, now, it is seen as a socioeconomic process depending on its knowledge content and effects (Rosenberg 1984).

According to the recent approach, technology and knowledge are often intertwined. Knowledge is regarded as the outcome of the research and innovation process, which requires complex and high-cost systems of "learning from each other." Consequently, technology transfer mainly depends on this complex and costly learning process (Levine et al. 1991; Kranzberg 1986). As a result of the above-mentioned changes, value generation has been shifting from physical and tangible issues to intangible assets. Management of learning and knowledge production processes are going to be more complex than ever.

There are new modes of knowledge production and knowledge transfer. These new modes begin with the definition of the problem, followed by problem solving, and end when the society is satisfied with the results from knowledge transfer and the ensuing production. In contrast to the old and traditional linear system, recent nonlinear or evolutionary knowledge production mechanisms depend on large networking, trans-disciplinary approaches, and interaction with the expectations of the society. All beneficiaries such as firms, public authorities, universities, R&D suppliers, society representatives, financers, etc., are involved and take part in these large networks (Nowotny et al. 2001).

M. Kiper (✉)
Technology Development Foundation of Turkey (TTGV), Ankara, Turkey
e-mail: mkiper@ttgv.org.tr

M.A. Yülek and T.K. Taylor (eds.), *Designing Public Procurement Policy in Developing Countries*, DOI 10.1007/978-1-4614-1442-1_5,

Some models have been trying to explain the changes in the research system with its social and technical context. The most popular among such models are:

- The model of National System of Innovation (Freeman 1995; Nelson and Rosenberg 1993)
- The model of an emerging "Mode2" of the production of scientific knowledge (Gibbons et al. 1994; Nowotny et al. 2001)
- The model of a Triple Helix of university-industry-government relations (Etzkowitz and Leydesdorff 1995, 1998, 2000; Etzkowitz 2001)

Technology Transfer Concept

Technology transfer is defined as "the process by which technology, knowledge, and/or information developed in one organization, one area, or for one purpose is applied and utilized in another organization, in another area, or for another purpose" (US Department of Energy). It has been argued that technology transfer strategies are crucial in the development and sustainability of a competitive knowledge-based economy. There are several definitions for technology transfer, which depend on the type or form of the transfer processes (Table 5.1).

Some Classifications for Technology Transfer

The exact nature of a technology transfer activity is difficult to pin down, partly because the term has many different connotations. So, this activity can be classified in several ways from different perspectives. Some categorizations are given below.

According to Relations and Size

Some types of technology transfer that are commonly discussed depend on its effects, results, and related parties (Singh and Aggarwal 2010):

- *International technology transfer*: the transfer of technologies developed in one country to firms or other organizations in another country.
- *North–south technology transfer*: activities for the transfer of technologies from industrial nations (the North) to less-developed countries (the South), usually for the purpose of accelerating economic and industrial development in the world.
- *Private technology transfer*: the sale or other transfer of a technology from one company to another.
- *Public-private technology transfer*: the transfer of technology from universities or government laboratories to companies.

Most technology transfer takes place because the organization in which a technology is developed is different from the organization that brings the technology to the market. The process of introducing a technology into the marketplace is called

Table 5.1 Some technology transfer definitions

- The process of sharing of skills, knowledge, technologies, methods of manufacturing, samples of manufacturing, and facilities among governments and other institutions to ensure that scientific and technological developments are accessible to a wider range of users. http://en.wikipedia.org/wiki/Technology_transfer
- The communication or transmission of a technology from one country to another. This may be accomplished in a variety of ways, ranging from deliberate licensing to reverse engineering. http://www-personal.umich.edu/~alandear/glossary/t.html
- The ability to take a concept from outside the organization (typically from a government or university research program) and create a product from it (Process). http://ccs.mit.edu/21c/iokey.html
- Exchange or sharing of knowledge, skills, processes, or technologies across different organizations. http://www.nsf.gov/statistics/seind06/c4/c4g.html
- The process of transferring scientific findings from research laboratories to the commercial sector. http://www.insme.org/page.asp
- The transfer of technology or know-how between organizations through licensing or marketing agreements, codevelopment arrangements, training or the exchange of personnel. http://www.et.teiath.gr/tempus/glossary.asp
- The transfer of knowledge generated and developed in one place to another, where it is used to achieve some practical end. http://www.rmauduit.com/glossary-t.html
- The process by which technologies developed by industrialized nations are made available to less industrialized nations. http://www.climatephilanthropists.org/basics
- The transfer of discoveries made by basic research institutions, such as universities and government laboratories, to the commercial sector for development into useful products and services. http://biotech-u.com/courses/mod/glossary/view.php
- The intentional communication (sharing) of knowledge, expertise, facilities, equipment, and other resources for application to military and nonmilitary systems. http://www.onr.navy.mil/sci_tech/3t/transition/tech_tran/orta/glossary/
- The diffusion of practical knowledge or the movement of modern or scientific methods of production or distribution from one enterprise, institution or country to another. Technology may be transferred by giving it away (e.g., through technical journals or conferences); by theft (e.g., industrial espionage); or by commercial transactions (e.g., licensing of patents for industrial processes, technical assistance, or training.). http://www.itcdonline.com/introduction/glossary2_q-z.html-trade-t.cfm

technology commercialization. In many cases, technology commercialization is carried out by a single firm. The firm's employees invent the technology, develop it into a commercial product or process, and sell it to customers.

From a business perspective: companies engage in technology transfer for a number of reasons:

- Companies look to transfer technologies from other organizations because it may be cheaper, faster, and easier to develop products or processes based on a technology someone else has invented rather than to start from scratch. Transferring technology may also be necessary to avoid a patent infringement lawsuit. Another aim may be to make that technology available as an option for future technology development. Finally, transfer is made in order to acquire a technology that is necessary for successfully commercializing a technology the company already possesses.
- Companies look to transfer technologies to other organizations as a potential source of revenue, to create a new industry standard or to partner with a firm that has the resources or complementary assets needed to commercialize the technology.

According to the Process

Technology can be transferred directly or indirectly depending on the process. From this perspective, license agreements, joint ventures, turnkey solutions, and consultant agreements would be direct transfer while transfer of technology via business visits, seminars, and training could be assumed as an indirect.

According to Absorption Capacity of Knowledge Buried in Technology

From the relation mentioned in group 2, especially from the recipients' perspective, technology transfer process can be classified as active or passive, depending on the assimilation capability of the tacit knowledge in the transferred technology, since most of the survey indicates that most of the knowledge available in the transferred technology is tacit and difficult to codify as a blueprint.

The term "tacit knowledge" is that which is embedded in the product, process, organization or individuals, and is difficult to imitate and transfer. It can be gained by deep involvement of the process. It depends highly on practice, is highly pragmatic, and can only be transferred by very effective sharing experience through special mechanisms. Facilitating the efficient mechanisms to reach and exploit tacit knowledge is one of the main factors of technology transfer. The diffusion, exploitation, and assimilation of tacit knowledge, which is embedded in a technology, is a highly critical factor, not only for firms but also for national Research Technology Development and Innovation (RTDI) policies from the macroperspective. When tacit knowledge infused in the technology is assimilated, the dependence on other countries for this technology is eliminated. Moreover, some of the most important aspects of national strategies are dissemination, exploitation plans, and activities, thus preventing continuous technology dependence to the outside.

According to National Policy Basis

Technology transfer concerns not only the people who pay for it, but the general public, especially because of:

- Criteria for its environmental effects (emission, rehabilitation cost for waste treatment, acceptance by society, etc.)
- Criteria for its qualitative base (novelty, market saturation, reliability, accesability, etc.)
- Criteria for its economical effects (payback period, adaptation time, sustainability, contribution to competition advantages-price, productivity, etc.)
- Criteria for its societal effects (public benefit, employment, human resource qualification, positive effects on other sectors or area, etc.)

Additionally, technology transfer, for example, is not about selling hardware to a client who is, then, left with the task of using it how he deems fit. Technology transfer is the imparting of knowledge, skills, and methodologies involved in the whole production cycle. Technology transfer is a system that encompasses the social and economic fabric of a country. Where technology has been effectively transferred, there should be a visible change from the person to the production system. Such change should be in compatibility with the needs and take place in the institutional framework, skills, training, financial capacity, promotion, and active support of endogenous capacity, and in appreciation of the natural environment of the recipient country. Technology transfer also has to do with disseminating information on the technologies themselves.

As mentioned before, the recent approach assumed that technology is derived from knowledge which is regarded as the outcome of a research and innovation process. This process requires large networking where all beneficiaries – firms, public authorities, universities, R&D suppliers, society representatives, financers, etc.– are involved. The relation between those parties are mainly horizontal, rather than vertical, since all beneficiaries want to access the output (tacit knowledge) resulting from RTDI activities to be used in line with individual expectations.

One of the meaningful technology transfer categorizations is horizontal vs. vertical transfer. However, in some cases, the vertical technology transfer concept is used for inner relation of the firms, while horizontal transfer refers to the firms' relations with outside parties. Briefly, in this article the vertical technology transfer concept is used in reference to transfer systems that has not been able to access and exploit tacit knowledge. Meanwhile, horizontal transfer terminology is used for technology handling by being an active participant in the RTDI networks.

From the above-mentioned perspectives some of the vertical transfer mechanisms are:

- Purchasing machinery, equipment, etc.
- Licensing
- Franchising
- Direct foreign investment
- Turnkey installation
- Joint ventures
- Procurement contracts
- Consultancy agreements
- Business visits

Meanwhile, the main horizontal transfer mechanisms would be as follows:

- Scientific and technical personal exchange
- Cooperative research
- Clustering and networking for RTDI
- Conferences and fairs
- Open literature
- Training
- R&D projects
- University–industry cooperation
- State aid programs to support RTDI projects

Obviously, there is not a black–white distinction between horizontal and vertical transfer systems. The key word for clarification of this classification is "absorptive capacity," which is the most critical point for technological learning.

Absorptive Capacity Framework

Technological capability is acquired through the process of technological learning. Effective technological learning requires absorptive capacity, which has two important elements: the existing knowledge base and the intensity of effort (Cohen and Levinthal 1990). First, the existing knowledge is an essential element in technological learning, as knowledge today influences learning processes and the nature of learning to create increased knowledge in the future. Accumulating the existing knowledge increases the ability to make sense of, assimilate, and use new knowledge. The foreign supplier may take an active role, exercising significant control over the absorptive capacity, into which the technology is transferred and used by the local recipient. Alternatively, the supplier may take a passive role, having almost nothing to do with the way the user takes advantage of the available technical know-how, either embodied in or disembodied from tangible artifacts.

This means that firms in developing countries have many alternative mechanisms for acquiring foreign technology, of which foreign direct investment, foreign licensing, and turnkey plants are the major sources of formal technology transfer (Hoekman et al. 2004). Contracted research with local universities and government research institutes becomes an important source of the industrialization process in the developing countries (OECD ST&I Outlook 2006). The purchase of capital goods transfers machine-embodied technology. Foreign suppliers and original equipment manufacturer (OEM) buyers often transfer critical knowledge to producers to ensure that the producers' products meet the buyers' technical specifications. Printed information, such as sales catalogs, blueprints, technical specifications, trade journals, and other publications, together with the observation of foreign plants, and even reverse engineering and industrial espionage, serve as important informal sources of new knowledge for firms in developing countries. In addition, reverse brain-drain or return of native foreign-trained professionals and moonlighting foreign engineers give significant rise to technological learning of the firm in the developing countries. If firms in developing countries have absorptive capability, they can effectively acquire foreign technology informally without any transaction costs.

Technology Trajectory Framework for Developed and Catching-Up Countries

The major steps of technology transfer and commercialization involve the transfer of

1. Technology codified and embodied in tangible artifacts
2. Processes for implementing technology
3. Knowledge and skills that provide the basis for technology and process development

This framework analyzes and integrates two technological trajectories: one in advance (developed) countries and the other in developing countries. Technological trajectory refers to the evolutionary direction of technological advances that are observable across industries.

Technology Trajectory in Advanced Countries

According to Utterback (1994), industries and firms in the advanced countries develop along a technology trajectory made up of three stages: fluid, transition, and specific (Utterback 1994).

Firms in a new technology area will exhibit a fluid pattern of innovation. The rate of radical (rather than incremental) product innovation is high. The new product technology is often crude, expensive, and unreliable, but it fulfills a function in a way that satisfies some market niches. Product changes are as frequent as market changes, so the production system remains fluid, and the organization needs a flexible structure to respond quickly and effectively to changes in the market and the technology (Abernathy and Utterback 1978; Utterback 1994).

As market needs are better understood and alternative product technologies converge or drop out, a transition begins toward a dominant product design and mass production methods, which adds competition in price and product performance. Cost competition leads to radical changes in production processes, which drive costs rapidly down. Production capability and output now assume greater importance to reap scale economies.

As the industry and its market mature and price competition grows more intense, the production process becomes more automated, integrated, system-like, specific, and rigid, thus turning out a highly standardized product. The focus of innovation shifts to incremental process improvements, seeking greater efficiency.

When the industry reaches this stage, firms are less likely to undertake R&D activities that are aimed at radical innovations, because the firms are more vulnerable in their competitive position. Industry dynamism may become regenerated through invasions by radical innovations introduced by new entrants (Anderson and Tushman 1990; Cooper and Schendel 1976; Utterback and Kim 1985). Often these are innovations generated elsewhere that migrate into the industry. Some industries, however, are quite successful in extending the life of their products in this specific state with a series of incremental innovations to add new values. It is at the later part of this stage that industries are typically relocated to developing countries, where production costs are lower. The upper part of Fig. 5.1 depicts the above model. This technology trajectory model may change significantly with a shift in the techno-economic paradigm (Freeman and Perez 1988). The Utterback model, however, is still useful in analyzing technology management issues in developing countries.

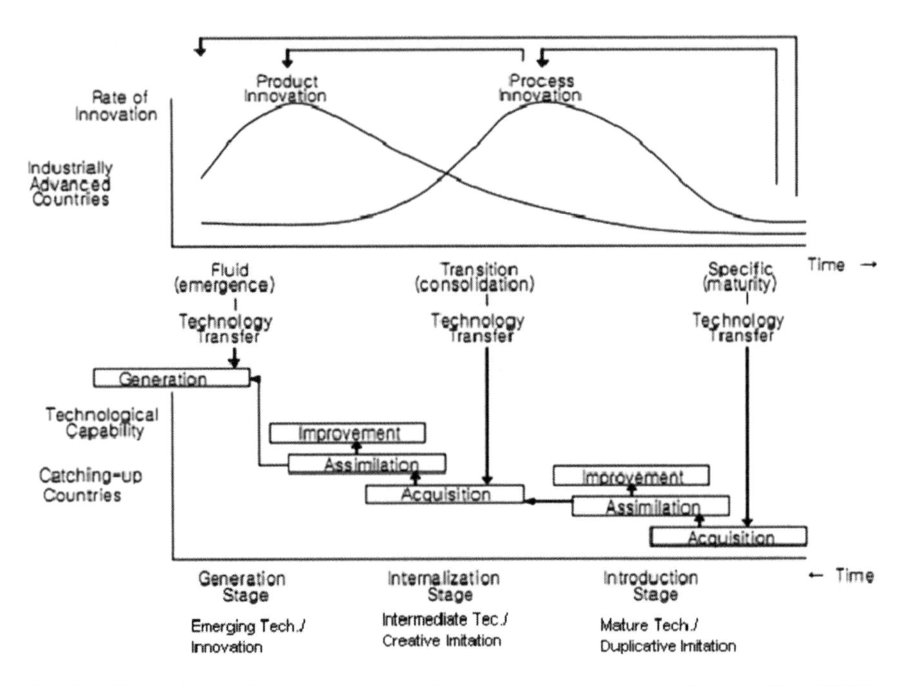

Fig. 5.1 Technology trajectory for developed and catching-up countries. Sources: Kim (1997), Lee et al. (1988)

Technology Trajectory in Catching-Up (Developing) Countries

On the basis of research for recently developed countries, for example Korea, Linsu Kim developed a three-stage model: acquisition, assimilation, and improvement to extend Utterback model (see Fig. 5.1).

At the early stage of industrialization, developing countries acquire mature (specific state) foreign technologies from industrially advanced countries. Lacking local capability to establish production operations, local entrepreneurs develop production processes through the acquisition of "packaged" foreign technology, which includes assembly processes, product specifications, production know-how, technical personnel, and components and parts. Production at this stage is merely an assembly operation of foreign inputs to produce fairly standard, undifferentiated products. For this purpose, only engineering efforts are required.

Once the implementation task is accomplished, production and product design technologies are quickly diffused within the country. Increased competition from new entrants spurs indigenous technical efforts in the assimilation of foreign technologies; in order to produce differentiated products. Technical emphasis is placed on engineering and limited development, rather than research.

The relatively successful assimilation of general production technology and increased emphasis upon export promotion, together with the increased capability of local scientific and engineering personnel, leads to a gradual improvement of mature technology. Imported technologies are applied to different product lines through local efforts in research, development, and engineering.

Integration of the Two Trajectories

Linking the technology trajectories of Utterback and Kim, Lee, et al. (Utterback 1994; Kim 1980; Lee et al. 1988) postulate that the three-stage technology trajectory in developing countries takes place not only in mature technology at the specific stage but also in intermediate technologies at the transition stage. As shown in Fig. 5.1, firms in developing countries, which have successfully acquired, assimilated, and sometimes improved, mature foreign technologies, may aim to repeat the process with higher-level technologies at the transition stage in the advanced countries. Many industries in the first tier developing countries (e.g., Taiwan, China, and Korea) have arrived at this stage. Some successful Korean industries have accumulated enough indigenous technological capability to generate emerging technologies at the fluid stage and challenge firms in the advanced countries (Kim 2000). Innovation is the watchword in these industries. When a substantial number of industries reach this stage, the country may be considered an advanced country. In other words, as shown in the lower part of Fig. 5.1, developing countries reverse the direction of the technology trajectory of the advanced countries. Thus, they evolve from the mature technology stage (for duplicative imitation), to the intermediate technology stage (for creative imitation) and to the emerging technology stage (for innovation).

The government policies related to technology development may also be assessed from the technology flow perspective. This perspective is mainly concerned with three key sequences in the flow of technology from developed to developing countries: (a) transfer of foreign technology, (b) diffusion of imported technology and indigenous R&D to assimilate and improve the imported technology, and (c) to generate own technology. The first sequence involves technology transfer from abroad through such formal mechanisms as the FDI, the purchase of turnkey plants and machinery, foreign licenses, and technical services. Effective diffusion of imported technology within an industry and across industries is a second sequence in upgrading the technological capability of an economy. The third sequence involves local efforts to assimilate, adapt, and improve imported technology and, eventually, to develop one's own technology. These efforts are crucial in augmenting technology transfer and expediting the acquisition of a technological capability. Technology may be transferred to a firm from abroad or through local diffusion, but the ability to make effective use of it is not easy to transfer. This ability can only be acquired through indigenous technological effort.

The integrative framework presented in Fig. 5.1 can also be applied to technology strategies in the private sector. First, it is important for firms to have an effective strategy for the acquisition of foreign technology, the diffusion of imported technology within the firms and in-house R&D. Second, the firms also need to have an effective strategy in creating the demand of new technology in the market, developing supply (R&D) capability, and coupling the market demand with the R&D capability. Finally, such strategies need to change along the reversed evolution from the mature technology stage to the emerging technology stage.

"Knowledge Zone" of the Knowledge Economy

A knowledge-based economy can be defined as an economy directly based on the production, distribution, and use of knowledge. In the knowledge-based (or knowledge-driven) economy, the innovation approach and the system of innovation depends on very complex social networks. Such a new system is very different from the traditional model which we may call as the linear innovation system. At the same time, the dynamics of the knowledge production and control system have also been changed sharply. Beginning from problem definition, the main tendency has been changing from the discipliner approach to the trans-discipliner one, in parallel with the changes in the innovation system. Similarly, universities and other research organizations have been playing more active and enhanced roles as R&D suppliers, and their relations with industry exhibit more highly complex dynamics than ever before in knowledge-based economy.

The above changes have been proposed as the recent models for the evaluation of the knowledge-based economic relations or driving forces of knowledge production and innovation systems.

As mentioned before, according to the recent approach, technology is derived from knowledge, and knowledge is the outcome of the Research, Technology Development and Innovation (RTDI) process. This process requires somewhat larger networking in that all beneficiaries – firms, public authorities, universities, R&D suppliers, society representatives, financers, etc. – are involved.

Conceptually, the main sequential RTDI steps are as follows:

Basic R&D: when new knowledge is discovered and general research must be conducted for further, basic understanding.

Exploratory R&D: when R&D is performed to uncover unknown applications of a technology.

Applied and Information-based R&D: when research is performed to apply basic knowledge in order to solve a particular problem.

Technology Development: when the R&D is "hardware" oriented and the application of research is already well thought out and designed.

After the RTD activities, the market penetration effort begins in order to put the RTD deliverables into particular markets as an *innovative output*.

The problems related with the public welfare, solution, implementation, formulation of regulations, and the use of outputs are all considered together. For this reason, industrial firms and their associations constitute product developers and producers, user associations, and regulation bodies, which represent infrastructures in hyper and complex systems.

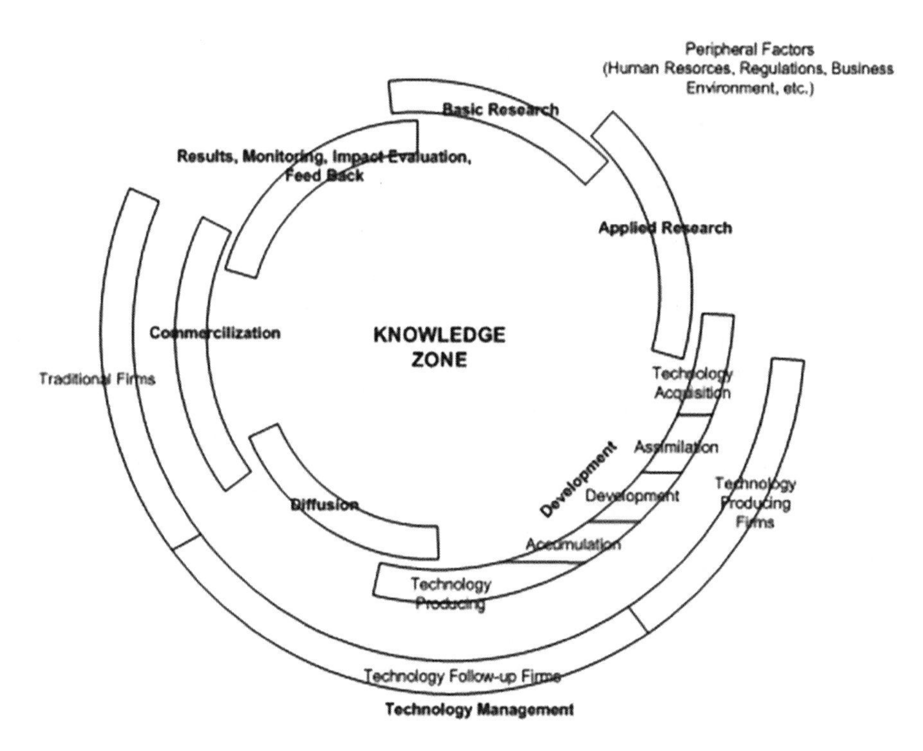

Fig. 5.2 Knowledge zone. *Source*: Kiper (2004)

Based on the features described above, recent technology production, development, and transfer systems in this area encompass the *Knowledge Zone* (Fig. 5.2). It may begin with basic research extended up to diffusion, commercialization, and performance measurement. As a result of complex and chaotic relations, the knowledge value chain is used to construct the innovation system base, which has been addressed from problem definition to solution utility, dissemination, and evaluation of hyper research working results.

As shown in Fig. 5.2, it is critical to:

- Produce the knowledge, assimilation, and diffusion of tacit knowledge as an output of technology development nationally and regionally. Therefore, technology dependence from the outside is prevented.
- Use this hyper and complex system chain for generating more high-tech start-ups and spin-off companies and also use this system for upgrading the firms from traditional companies to technology followers and then to technology producing leaders.
- Design continuous and systematic evaluation, since analysis of effects and feedback provision play an important role in the chain. It is expected that the system should be improved continuously.
- Establish a proper consultancy and management mechanism to be used in the application of this complex and chaotic cycle at each level of the hyper system.
- Put an effort to create many different methods for bringing all the parties together in order to activate the application tools of this cycle.

Since commercial firms seek to use knowledge and innovation as a result of R&D network in a global scale, the development of an international R&D structure becomes the central issue (Pack and Westphal 1986). Zanfei describes the new organizational mode of transnational innovation as a "double network" comprising the internal and external networks (Zanfei 2000). The internal networks refer to the inner mechanisms, while external networks are constituted by relations with actors outside to the organization.

While universities and other public and private research organizations act as R&D suppliers, firms are the main drivers for putting the research results into the market and for improving the welfare of the society by commercializing these results within the economy.

Firms may jump to some high ability level with technology transfer, assimilation, development, and production loop, which are naturally generated in these processes, and firms gain absorptive capacity for further stages. Due to these functions, firms can be assumed as one of the main forces of the "Knowledge Zone." Firms can be grouped into three categories depending on their ability to take part and play their role in such research networks, their capacity to assimilate and commercialize research results and their market position.

1. *Traditional Firms*: These firms produce noncomplex products for traditional sectors with small production scales. Their competitive capabilities are mainly dependent on low cost and flexible production.
2. *Technology Follower Firms*: Their production systems depend on automated machines and flexible work units. The production strategy of these firms is to produce new products in small batches within short production periods using TQM or other advanced management techniques. Inter-firm relations depend on Just in Time (JIT), production networks, service facilities, and stable main producer–subcontractor relations. They have trained manpower and multifunctional working cultures and make R&D investments at moderate levels with low risk.
3. *Technology Producer (Leader) Firms*: These firms are the industry leaders and take the first action. They undertake R&D investments at high levels with high risk. They create technical barriers against their followers or competitors with standards and technical regulations, and protect their products by patenting or by other IPR mechanisms.

Technological innovation may be the main source of high market shares. This is due to the different specialties of products, which may also lead to a unique application and/or a first sample of its typology. This kind of innovation is providing exponential profit margins. At the same time, the new or reshaped harmonized technical specification, which provides a long-term competition source that generally results in technological innovation (Fig. 5.3).

The technology producing firms can be placed in the first section. The followers can be placed in the second section and the traditional firms are placed in the third section of Fig. 5.3. For this reason, technology production and development becomes one of the major policy issues, not just at the microlevel, but at the regional or even national level.

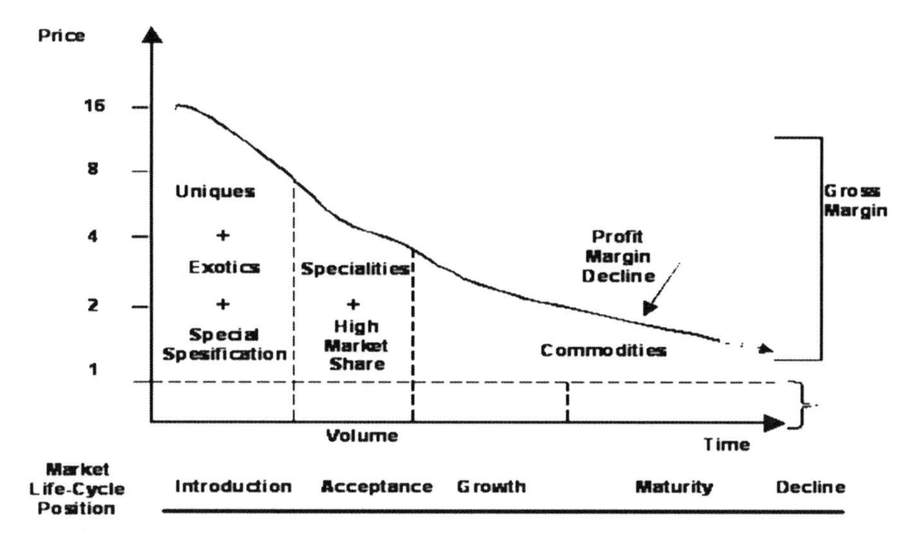

Fig. 5.3 Technology application spectrum (TAS). *Source*: Hruby F.Michael, Technoleverage, AMACOM (1999)

Before the dissemination of the results, depending on the special features of outputs, another important subject, which should be regarded in the "Knowledge Zone," is to make a normalization process by standardization or through technical regulation activities in the particular area. As a result of these research works beginning from the design phase, all works should be carried out in close contact with organizations that represent the society, since markets and end-users are deeply affected from the regulation of the specific field by rules, norms, laws, etc.

At the same time, it is necessary to be ready against the effects of technical regulations, which most probably will be one of the major factors of competition between firms and nations. Together with dissemination, commercialization is another important process for nations who support the RTDI activities and expect to benefit from outputs, as well as other stakeholders and financers of the research works. Impacts of the commercialization process should be assessed carefully. Evaluation of all the results generated from each step should be monitored and its impact should be carefully assessed to get valuable data, not only for reshaping and replanning of the remaining activities, but also for new research inputs.

In Fig. 5.4, some relations between the elements of the "Knowledge Zone" and their positions relative to different scales (macro, mezo, micro) are presented. In this figure, the macro scale can be assumed as a national or transnational level, the mezo scale can be defined in a regional or sectoral level within the country and the micro scale can be taken as a firm level. Although there is no sharp distinction about which element is directly positioning into which scale, with regard to the functions of the elements, basic research, some dissemination activities (especially horizontal) and some monitoring, assessment and results, which effect the macro policies at the national or even greater levels can be linked to the macro scale.

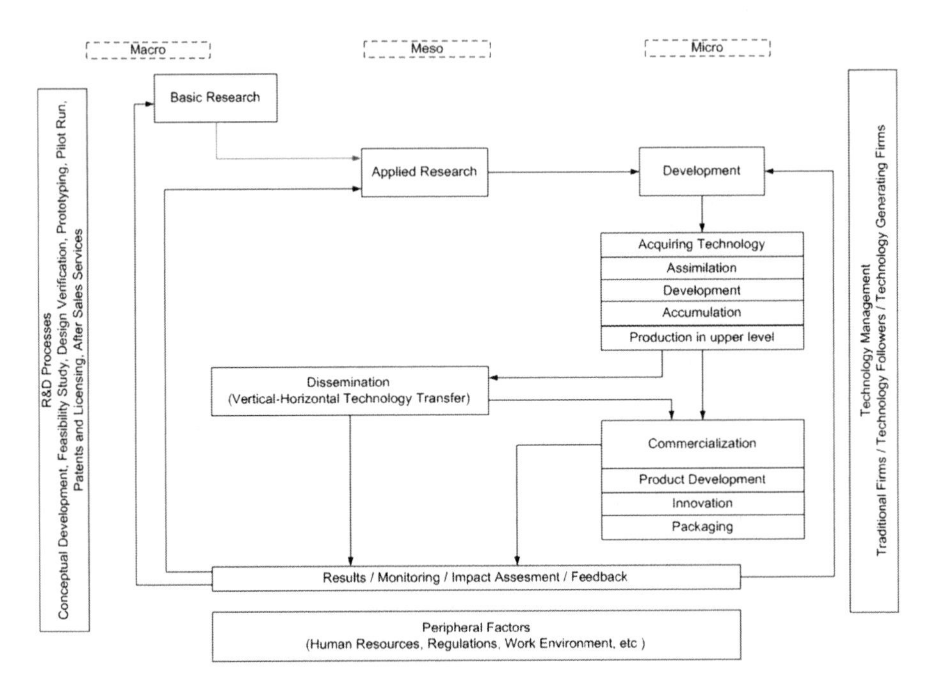

Fig. 5.4 Some relations between elements of knowledge zone. *Source*: Kiper (2004)

Applied research, some dissemination, and monitoring activities, impact assessment and results which are sectoral or locally oriented can be related to the mezo level, while outputs of research acquired by firms, can be assumed to be at the micro level.

Conclusions

A knowledge-based economy depends on the knowledge production capacity and the ability to adapt it to new or developed technologies. Some aspects of knowledge become codified and highly mobile at a global scale, while other key elements remain tacit and deeply embedded in individuals, organizations, and localities.

Technology and technology transfer strategies are more critical than ever because of their impact on national, regional, and even the global scale, and the fact that technological advance has been recognized as the key driving force for economic development. Thus, for the developing countries, it is the only way to place the acquisition and progressive mastering of technologies at the center of national policies in order to decrease the technological gap with and prevent continuous dependence on the developed countries.

Successful transfer always necessitates capacity. Countries are able to acquire international technology more readily if domestic firms have local R&D programs, there are domestic private and public research laboratories and universities, and there exists a sound basis for technical skills and human capital (OECD

ST&I Outlook 2006). All this reduces time and the costs of assimilation, creative imitation, and follow-on innovation.

Previous experiences showed that copying from other national policies does not work. A one-size-fits-all approach to policy will clearly be inappropriate, but some lessons could be derived from the strategies applied in the recently developed countries. One of the critical issues in this sense is how to handle and design the national IPR regime.

Japan is a preeminent example of a country that developed technological capacity rapidly. One reason for Japan's rapid growth and industrialization after World War II was that its patent system was designed for both small-scale innovation and diffusion. The system encouraged incremental and adaptive innovation by Japanese firms and promoted the diffusion of knowledge, including foreign technologies, into the wider economy (Maskus 2000).

Korea is another technology follower that encouraged learning via duplicative imitation of mature technologies that foreign firms had permitted into the public domain or were willing to provide cheaply (Kim 2000). IPRs were weak and they encouraged imitation and adaptation during this stage. The government essentially promoted the development of technical skills through education and workplace training and ensured the absence of an antiexport bias. In the 1980s, Korea shifted to creative imitation, involving a more significant transformation of imported technologies. This required domestic R&D and in-house research capabilities to adapt technology. The government became more welcoming to formal channels of TT and strengthened the IPR regime when Korea had reached a technologically competent level.

Brazil, Malaysia, and the export-intensive regions of China and India, are other examples of movement from duplicative to creative imitation. In these cases, IPR protection was initially limited and firms took advantage of the available foreign technologies. As the technological sophistication of production processes matured and the depth and complexity of knowledge for effective absorption grew, firms increasingly resorted to formal means of TT, and the governments strengthened the IPR regime.

Many middle-income developing countries are at the duplicative imitation stage, hoping to absorb free or cheap foreign technologies into labor-intensive export production and evolve higher value-added strategies over time. In order to shift from duplicative to creative imitation and, later, to come to the capacity for emerging technology at global level, it is critical to establish and initiate a "Knowledge Zone," as described earlier.

Learning and also innovation are interactive processes and they depend on trust and other elements of social cohesion (Lundvall and Tomlinson 2001). These are crucial factors for nonlinear, evolutionary systems, and because of their chaotic, cultural, and time-dependent behavior, it is not easy to create "trust" in a short time. This paradox can be solved within a well-defined and planned organizational structure and among the participants who know each other well. Such social and behavioral concerns should also be considered in the flexible "Knowledge Zone."

Expected skills for the work force will be of highest quality than ever and their qualifications will not only depend on technical acumen but will also require some other special features, such as communication ability and advanced management skills. Consequently, education and training of the workforce should generate the required management and technical skills.

Technology transfer cannot be handled as a separate issue from national techno-economical perspectives. It has direct or indirect relations with almost all national innovation system parameters.

References

Abernathy WJ, Utterback JM (1978) Patterns of industrial innovation. Technol Rev 80(7):40–47

Anderson P, Tushman M (1990) Technological discontinuities and dominant designs: a cyclical model of technological change. Admin Sci Q 35:604–633

Cohen WM, Levinthal DA (1990) Absorptive capacity: a new perspective on learning and innovation. Admin Sci Q 35:128–152

Cooper A, Schendel D (1976) Strategic responses to technological threat. Business Horizons 19:61–69

Etzkowitz H (2001) The second academic revolution and the rise of entrepreneurial science. IEEE Technol Soc 22(2):18–29

Etzkowitz H, Leydesdorff L (1995) The triple helix – university-industry-government relations: a laboratory for knowledge based economic development. EASST Rev 14(1):9–14

Etzkowitz H, Leydesdorff L (1998) The Triple Helix as a model for innovation studies. Sci. Pub. Policy 25(3):195–203

Etzkowitz H, Leydesdorff L (2000) The dynamics of innovation: from national systems and "Mode 2" to a triple helix of university-industry-government relations. Res Policy 29 (2):109–123

Freeman C (1995) The national system of innovation in historical perspective. Cambridge J Econ 19:5–24

Freeman C, Perez C (1988) Structural crises of adjustment, business cycles, and investment behavior. In: Dosi G, Freeman C, Nelson R, Soete GSL (eds) Technical change and economic theory. Pinter Publishers, London, pp 38–66

Gibbons M, Limoges C, Nowotny H, Schwartzman S, Scott P, Trow M (1994) The new production of knowledge: the dynamics of science and research in contemporary societies. Sage, London

Hoekman BM, Maskus KE, Saggil K (2004) Transfer of technology to developing countries: unilateral and multilateral policy options. Working Paper, Institute of Behavioral Science, University of Colorado, Colorado

Hruby FM (1999) TechnoLeverage: using the power of technology to outperform the competition. AMACOM Books, New York

Kim L (1980) Stages of development of industrial technology in a developing country: a model. Res Policy 9:254–277

Kim L (1997) Imitation to innovation: the dynamics of Korea's technological learning. Harvard Business School Press, Boston

Kim L (2000) Technology and industrial development: analytical frameworks. Korea Government Reform Council, Korea

Kiper M (2004) Teknoloji Transfer Mekanizmaları ve Bu Kapsamda Üniversite-Sanayi İşbirliği. in *Teknoloji Kitabı*. TMMOB, Ankara

Kranzberg M (1986) The technical elements in international technology transfer: historical perspectives. In: McIntyre JR, Papp DS (eds) The political economy of international technology transfer. Quorum Books, New York, pp 31–46

Lee J, Bae Z, Choi D (1988) Technology development processes: a model for a developing country with a global perspective. R&D Manage 18:235–250

Levine MD, Gadgil A, Myers S, Sathaye J, Stafurik J, Wilbanks J (1991) Energy efficiency, developing nations, and Eastern Europe. Lawrence Berkeley Laboratory, Berkeley

Lundvall BÅ, Tomlinson M (2001) Policy learning by benchmarking national systems of competence building and innovation. In: Sweeney GP (ed) Innovation, economic progress and quality of life. Edward Elgar Publishing, Cheltenham

Maskus KE (2000) Intellectual property rights in the global economy. Institute for International Economics, Washington

Methodological and Technological Issues in Technology Transfer: http://www.grida.no/publications/other/ipcc_sr/?src=/climate/ipcc/tectran

Nelson R, Rosenberg N (1993) National innovation systems: a comparative analysis. Oxford University Press, New York

Nowotny H, Scott P, Gibbons M (2001) Re-thinking science: Knowledge and the public in an age of uncertainty. Cambridge: Polity Press

Pack H, Westphal LE (1986) Industrial strategy and technological change. J Develop Econ 4:205–237

Rosenberg N (1984) The science/technology relationship, the craft of experimental science, and policy for the improvement of high technology innovation. Res Policy 13(l):3–20

Singh A, Aggarwal G (2010) Technology Transfer Introduction, Facts & Models. International Journal of Pharma World Research, 1(2)

OECD Science, Technology and Industry Outlook (2006)

Utterback JM (1994) Mastering the dynamics of innovation. Harvard Business School Press, Boston

Utterback JM, Kim L (1985) Invasion of stable business by radical innovations. In: Kleindorfer PR (ed) Management of productivity and technology in manufacturing. Plenum Press, New York, pp 113–151

Zanfei A (2000) Transnational firms and the changing organization of innovative activities. Cambridge J Econ 24(5):515–543

Part II
Country Experience

Chapter 6
The Complexities of Development: The South African National Industrial Participation Programme in Perspective

Richard Haines

Introduction

From the late 1990s, the National Industrial Participation Programme (NIPP) and the associated Defence Industrial Participation (DIP) programme have become one of the more visible development-oriented programmes of the post-apartheid, African National Congress (ANC), government. Industrial participation and countertrade have remained surprisingly resilient as methods of public procurement exercises and potential means of leveraging technology transfers and economic and industrial development. However, analyzing the impact of such programmes and deciphering their myths (Balakrishnan and Matthews 2009) are challenges to the relevant state agencies, industry practitioners and scholars alike. Andrea Hurst points out that development analysis and prescription should be grounded in a greater conceptual appreciation of the complexities and contradictions of social and economic reality (Hurst 2010).

This chapter comprises four aspects. First, a selective overview of the NIPP undertaking, in regard to both its recent history and current development, is provided, with a special reference to the defence offset aspects. Second, the emergence and content of the Competitive Supplier Development Programme (CSDP) – a partial alternative to NIPP – is discussed. Third, the dynamics of the impact of the Strategic Defence Programme (SDP) and the associated DIP on the South African defence industrial base are examined. Finally, a set of compact conclusions and policy recommendations are presented.

R. Haines (✉)
Nelson Mandela Metropolitan University, Port Elizabeth, South Africa
e-mail: richard.haines@nmmu.ac.za

M.A. Yülek and T.K. Taylor (eds.), *Designing Public Procurement Policy in Developing Countries*, DOI 10.1007/978-1-4614-1442-1_6,

The National Industrial Participation Programme in Perspective

Contextual History of the NIPP

Since its formal inception during the later 1990s, the NIPP has been one of the more visible and publicly debated aspects of South Africa's industrial policy. This is primarily due to the strong relationship between industrial participation and defence procurement, especially in regard to the controversial SDP, the so-called "arms deal" in local and international parlance. When the South African Parliament decided on a major upgrade of the South African National Defence Force (SANDF) weaponry – fighter planes, jet trainers, helicopters, submarines and corvettes – the issue of offsets quickly assumed prominence.

DIP was originally instituted by Armscor during 1988 and until 1996 the countertrade programme entailed direct offsets, indirect offsets and countertrade, which mainly existed in the form of counter-purchased goods for export. Although informal arrangements can be found before the mid-1990s in the non-defence field, the current NIPP exercise can be traced primarily to the restructuring of economic and industrial strategy and policy in the early and mid-1990s.

In South Africa the new industrial participation strategy was one of the pillars of the economic and industrial policy interventions unveiled during 1996–1997. Incorporating policy dialogue with the World Bank in Washington (see, e.g. Bond 2002), the Growth, Employment and Redistribution (GEAR) strategy formed the touchstone for supportive national industrial, regional and spatial development initiatives (SDIs) (RSA Government 1996a, b).

However, the current breadth, depth and formalization of industrial cooperative interventions in the national economy need to be appreciated against a set of overlapping discourses and developments both within and without South Africa. Internationally, by the early 1990s, informed by the so-called "Washington Consensus", the advocacy for transitional and developing economies, including former state socialist economies, was for greater financial liberalization, reduced state spending, a shift to more market-oriented models and the expansion of export-led industrialization and trade. This meant, *inter alia*, reducing tariffs and explicit incentives and looking for reductions in the subsidization of industries. In tandem with these developments, the decline in defence spending in the wake of the Cold War and the changing nature of defence industrial production facilitated more complex sets of practices and interventions. This, in turn, linked procurement of defence and also non-defence equipment and services more directly with industrial cooperation, offsets and countertrade options (Dunne and Haines 2006).

In South Africa, under guidelines that took effect from September 1996, all government and parastatal contracts with an import content exceeding USD10 million were to include an Industrial Participation (IP) component. The value of the offsets was to comprise a minimum 30% of a bid's imported component for civilian contracts. For defence contracts, the offsets should comprise 50% of a bid's imported components (DTI (Department of Trade and Industry) 1997). The industrial participation portion of the bid was assessed according to "credits" awarded for each type of benefit. Bidders in general were to fulfill their obligations within 7 years and provide a performance guarantee equal to 5% of the offset component. At the time South Africa was the only country to utilize a dual approach to offsets.

In regard to the SDP, all contracts with a value of greater than USD10 million were to be subject to the National Industrial Participation Policy, administered jointly by the Department of Trade and Industry (DTI) and the Department of Defence (DoD).

The SDP constitutes one of the pillars of the government's expanding set of Industrial Participation (IP) projects. The SDP includes both defence-related countertrade investment, namely the DIP scheme that is ultimately the responsibility of the DoD, and non-defence-related investment, namely the NIPP scheme, which falls under the DTI. The latter scheme is, in turn, subsumed under the overall NIPP exercise. Traditionally, in its reporting the DTI maintained a clear distinction between the defence and non-defence aspects, but over time this distinction has blurred. In part, this has been a response to the widespread criticism of the defence package and its defence and non-defence offsets. The DIP scheme can be further broken down into direct and indirect DIP. The indirect DIP is related solely to exports. It can include investment into creating capacity for export market or a technology transfer for export market, though these two scenarios did not materialize.

The NIPP programme is managed and administered by the DTI, while the DIP obligations are jointly managed and administered by Armscor and the DoD's Acquisition Division. In practice, the relevant directorate in Armscor has played the lead role in the management of the DIP obligations, with the DoD and Denel (South Africa's largest producer of defence equipment) playing supportive roles. The Armscor directorate responsible for the management of the offsets cooperates relatively closely with a specialist IP section in the DTI Office (Haines 2004).

The overall aim of the NIPP and the associated DIP programmes was "to raise investment levels and increase exports and market access for South African value-added goods and services by leveraging off government procurement" (DTI 2003, p. 2). Key objectives in original conception were as follows:

- To diversify and expand South Africa's industrial base
- To create jobs
- To focus investments in strategic areas
- To promote export-oriented ventures

Additional objectives were as follows:

- To ensure that there was appropriate technology transfer and R&D collaboration
- To help promote BBBEE (Broad-Based Black Economic Empowerment)

Target areas or sectoral emphases were clothing and textiles; auto, auto components and transport; agro-processing; mining, metals and minerals beneficiation; chemicals and biotechnology; crafts; and information and communications technology (Haines 2004).

In 2000, personnel at the South African ministries of Finance and of Trade and Industry estimated the overall return on the offsets associated with the SDP to be in the region of 94.5%, and that, during the duration of the deal, the anticipated exports

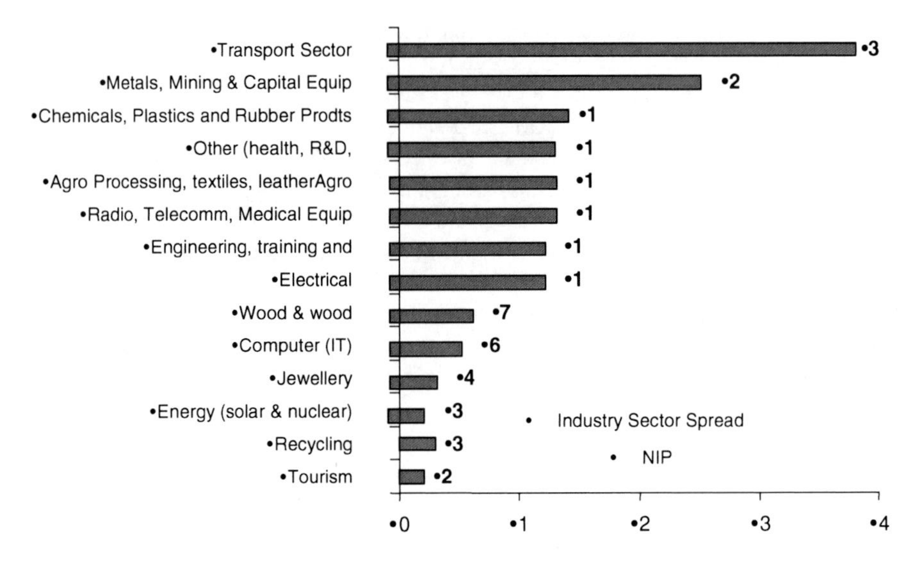

Fig. 6.1 The industry spread of 225 South African national industrial participation obligations projects over the period 1997–2008. *Source*: De Risi 2009

would be of the order of 280% of the original purchase price (Haines 2004, p. 303). Furthermore, according to its protagonists at the time, the SDP programme (NIPP and DIP aspects collectively) would generate some 65,000 jobs – a figure which was questioned repeatedly over the years by assorted critics.

National Industrial Participation Programme: Official Achievements

Despite a range of criticisms over the years, especially in regard to projects tied to the defence procurement, there has been a significant expansion and consolidation of the NIP programme during the 2000s. In late 2009, the DTI reported that obligations had been monitored in the region of USD 15.5 billion with the core obligations stemming from defence procurement and purchases by the South African Airways (SAA). Other obligations include transport, IT, telecoms and energy sectors (Eskom and PBMR). Two hundred and twenty-five projects were implemented in the period 1997–2008 with an industry spread as indicated in Fig. 6.1. Around USD 16 billion of credits have been awarded for projects during the period 1998–2008. Of this figure approximately USD 0.5 billion appears to be attributable to the Foreign Direct Investment (FDI) (De Risi 2009). In addition, the projects were distributed on a regional and provincial basis as follows: 113 projects in Gauteng, 27 in KwaZulu Natal (KZN), 4 in Limpopo, 5 in Mpumulanga, 4 in North West Province, 27 in Western Cape, 1 in Northern Cape, and 21 in Eastern Cape.

The DTI's 2009 performance review of the NIPP, which was tabled in September 2010 in Parliament, noted that by July, 221 projects and credits in excess of R19 billion had been awarded under the NIPP – both in investments made and exports generated. These projects had led to the creation and/or retention of 85,062 jobs, 26,442 direct and 58,620 indirect, boosted gross domestic product by R14.8 billion and contributed R11 billion to the balance of payments at 2008 prices.[1]

The official methodology was adapted with the experience gained from the NIPP. By the end of the 2000s, the DTI division charged with programme implementation had learnt several lessons in regard to IP methodology, in particular:

- Assistance to go beyond financial contribution
- Flexibility is essential
- Do not wait too long to realize a project, if it takes too long to close chances are it will not happen, or happen in time
- Guidelines are just guidelines, and are often tailored to suit a particular industry or situation (De Risi 2009)

Evaluating the National Industrial Participation Programme

While the official records reflect a healthy state in regard to the discharge of military and civilian obligations, there are less sanguine assessments and viewpoints. Estimating the real value of the NIPP and the associated DIP programme is a problematic exercise which raises issues of evaluating development initiatives within a complex socio-economic reality. Insufficient data is available for dispassionate analysis, and there are a range of hidden, contextual and transaction costs.

The on-going controversy and revelations over the SDP suggests that corruption has impacted the conceptualization and implementation of the initiative. Dr Gavin Woods, chairman at the time of parliament's standing committee on public accounts observed in 2001: "The big story of the arms deal ... was about how people in high places used the deal to enrich themselves at the expense of the South Africa public" (cited in Haines 2007).

The emerging body of critical international literature on the impact of offsets suggests the need for circumspection, and rigorous evaluation for feedback and policy adaption and modification. The cumulative impact of offsets is often problematic in terms of job creation, the strengthening of backward and forward linkages, and technology enhancement (Matthews and Williams 2000, pp. 26–31; Brauer and Dunne 2004; Dunne and Haines 2006). It has been argued that they do not constitute a "third way" for the economic development of developing countries (Matthews and Williams 2000; Batchelor and Dunne 2000). In a recent paper on the international experience of defence offsets, Jurgen Brauer and Paul Dunne argue

[1] *Business Day*, 14 September 2010.

that with substantial IP obligations, the likelihood is that the equipment purchased will come with a premium – often in the region of 30% on the off-the-shelf price. This premium or offset "overcost", they argue "does not buy general economic development, does not buy new and sustainable work, and, except for limited specific cases, does not result in appreciable technology transfer" (Brauer and Dunne 2010). However, in certain sectors, such as higher education, there is distinct scope for using the social capital and leverage that international firms can offer to build cost-effective partnerships and projects (Msomi and Shelver 2010).

Costs of the Equipment

The costs of equipment and services procured under industrial participation and countertrade agreements by the South African government is an area requiring significant future research. There is evidence ranging from anecdotal and circumstantial through to formal and direct that procurement with an IP condition to it often occurs at levels above market price or off-the-shelf prices (especially for defence equipment). The SAA recently disclosed to the South African Parliament's Public Enterprises Committee that it paid 3–5% above market price for its aircraft. The cause, the company stated, is the government's NIPP. Sandra Coetzee, SAA's Acting CEO, explained to the committee that suppliers build the cost of the offsets into the purchase price.[2] In regard to the SDP, it would appear that government in some instances paid distinctly more than prevailing prices for certain of the equipment.

The SDP ushered in a more complex approach to the evaluation of the bidding process, with a greater number of political and administrative stakeholders. The SDP included corvettes (with onboard helicopters), submarines, light jet trainers, jet fighters, and helicopters. Future procurement programmes with a special emphasis on the needs of land-based forces, and naval near-shore operations were envisaged. The early frontrunners with regard to the jet trainers and corvettes were respectively Aermacchi of Italy and Bazan of Spain. Bazan had been already been appointed preferred supplier for the acquisition process for corvettes begun in 1994. Subsequent rounds of negotiation saw the Spanish bid replaced by that of the German corvette consortium, following a visit by the then Vice President Thabo Mbeki to Germany (Haines 2007).

The price of the German ships was significantly higher but a key factor in this bid and the related German submarine consortium bid was a raft of impressive offset offers including the establishment of a metals beneficiation cluster in the planned Coega Industrial Development Zone (IDZ). The Aermacchi M 339 trainer was replaced by the Hawk jet trainer at a markedly higher price.

---

[2] *Draft CTO report*, September 2010.

Furthermore, as the actual cost of equipment was only about a third of the cost of the equipment cycle, the Aermacchi would have allowed the South African Air Force (SAAF) to trim its maintenance costs further given a relatively long-standing relationship between Italy and South Africa regarding the LIFT (Lead-in Fighter Training) aircraft. Furthermore, the Royal Australian and Canadian Air Forces both appear to have purchased the Hawk LIFT aircraft at significantly lower prices (Haines 2009a).

The acquisition of the new Gripen JAS-39 fighter aircraft was initiated in the same year (1997) as the previous generation of Cheetah C and the two seater Ds were taken into service by the SAAF. These Cheetahs were designed to stay in commission until at least 2017. South Africa subsequently attempted to sell a part of the stocks of the aircraft, much of which were essentially new, to certain governments. In 2009, for instance, Denel was in negotiation with Equador for the purchase of around 12 of the aircraft, promising a complete 5-year maintenance and support plan with an option of renewing this plan.

South Africa paid a good deal more for the Gripen than the Swedish air force. In addition, because they were the first non-NATO client, it could be argued that the country effectively paid for much of the development costs of the non-NATO base line version of the fighter (Haines 2009a).

According to local defence contractor Richard Young, the DoD's then Chief of Acquisitions stated in July 1999 that "we the Government do not care about the defence equipment itself; we are only interested in the counter-trade". By contrast, as the former Minister for Trade and Industries Alec Irwin reiterated in 2008 that the government "made it absolutely clear [in 2001] that the defence procurement, the equipment was the prime objective. That was the basis for making the decision, that was why we did it, and industrial participation and defence participation and the employment that may accrue from it, was a corollary of this, a benefit that we were attempting to obtain as we procured the defence equipment. It was not the reason for procuring the defence equipment".[3]

Considering Selected NIPP Projects

The official analyses of the NIPP exercise (defence and non-defence components) underline strongly the achievements. However, independent inquiry suggests more unevenness in the performance record of the obligors and the associated projects.

In the 2001 DTI hearings, the leading agencies for organized business and trade unions – the South African Chamber of Business (SACOB) and the Congress of South African Trade Unions (COSATU) – came out against offsets and the SDP more generally. SACOB argued that offsets were not likely to have any meaningful impact on economic development, and the COSATU questioned the priorities of

[3] Transcript of Strategic Briefing Issued by Government Communications (GCIS), 6 August 2008.

spending on new defence acquisitions, and highlighted the need for more specific efforts in regard to job creation. While the points of view of these two organizations have become somewhat muted and less explicit over time, this line of thinking has tended to persist among sections of their membership (Haines 2009a).

Selected studies in several provincial regions and nationally in the early and mid-2000s revealed the following set of findings:

- There was insufficient liaison with organized business and labour and local development agencies in centres such as Durban, Port Elizabeth and Cape Town in the identification, conceptualization and implementation of offsets (Haines 2004, 2005, pp. 59–73).
- In a few instances where data were available, the size of the investment and the claimed jobs created did not always tally with on the ground inquiry with relevant representatives of the management of the company in question (Haines and Wellman 2005, pp. 25–34).
- There was a relatively widespread perception among local private and public sector policy agents that the allocation of projects within the provinces was conditioned in part by political imperatives. There was a feeling expressed by a number of the interviewees that the selection of initiatives occurs more on a project-by-project basis with relevant consortia, than in a more integrated and systematic process. In such a process, the deployment and potential effects are assessed and contractors informed by the kinds of offsets they can arrange, and not always with detailed knowledge of local economic development requirements (Haines 2004, 2006; Haines and Wellman 2005, pp. 25–34).
- Assessment of the nature and impact of the NIPP projects during this period is further complicated by the fact that a number of offset investments and projects were channelled into brownfields enterprises. And in general obligors were able to maximize export credits available through the IP methodology through relatively modest but shrewd investments (Haines 2004, 2006; Haines and Wellman 2005, pp. 25–34).
- While there were problems with some of the more ambitious projects – to date those originally scheduled for the Coega IDZ and the British Aerospace (BAE) Systems-Harmony Gold-Mintek precious metals beneficiation venture – there were certain strategic investments. For instance, BAE's investments in the KZN biotech firms have proved vital in KZN's emerging biotech industry. A soft loan of USD 2 million for SA Bioproducts proved crucial for the survival of the firm, and has contributed in part to the creation of new production facilities and to the expansion of its domestic and export sales. Similarly, a BAE-IDC equity investment of R10 million (with a total of R23 million scheduled on performance) has enabled Biocontrol Products to expand its product base and assist in sales to the African market (Haines 2005, pp. 59–73).
- Projects tied up with obligors' core business were generally successful, e.g. Aerosud manufacturing interior components for Boeing (Haines 2004).
- Some of the interventions, though not particularly successful in terms of original intentions, had a positive unanticipated spin-off. An example here is the

refurbishment and expansion of Port Elizabeth's McArthur baths complex on the Humewood beachfront (Haines 2009a).

– There are a few modest but important investments in South Africa's maritime industry (Haines 2009b).

A study of the KZN (Haines 2005, pp. 59–73) found that, despite some modest but quite useful projects, there was little in the way of efforts in integrating these with the long-range development planning exercises for the city of Durban, such as that carried out by the International Monitor Group in the early 2000s (Monitor Company 2000). A further and key finding was that, despite formal requests made to the government, and especially the DTI, by Durban's surviving shipyard (Southern African Shipyards) for consideration for the DIP and the NIPP offset work, nothing was forthcoming. A growing marginalization in regard to government procurement contracts impacted adversely on the company's fortunes and contributed towards its subsequent closure. Moreover, although some procurement work in the maritime construction and repair field was lodged with smaller firms over time, South Africa effectively lost most of its competitive capacity to build ships of any real size.

A Western Cape regional study (Haines and Wellman 2005, pp. 25–34) found that the nature and impact of the selected NIP projects differed noticeably from official analyses at the time. One of the central findings was that high-growth emerging sectors, where offset investment would have been to the mutual benefit of both the local economy and obligors, were not identified and/or ignored. These sectors included the IT and software design and applications and boat and yacht building. The latter in particular had already gained a growing international market.

Studies regarding the conceptualization and implementation of offsets in Eastern Cape in the late 1990s and early and mid-2000s also found some tension between the claimed benefits of the projects and the experiences at local and projects level. Originally, the flagship NIPP offset applications within the Eastern Cape, and, indeed, of the defence-related NIPP programme nationally comprised a range of metal and metal beneficiation projects in the new Coega IDZ and the deep-water harbour some 22 km from Port Elizabeth. The centre-piece of the anticipated offsets was a steel mill which Ferrostaal was to facilitate. Batchelor and Dunne (2000, p. 22) argue that the promise of such offsets helped tip the decision to proceed with the IDZ at Coega. However, this and related offsets were not undertaken which has contributed to the under-performance of the CIDZ enterprise during the 2000s (e.g. Hosking and Haines 2005, pp. 1–23). Other alternative ventures were explored, such as a condom factory and beer brewing facility, without success. Thereafter, the obligor took over the management of the state-owned Magwa Tea Estate in North-Eastern Cape via a facilitating company. This, too, proved a problematic investment. The province saw some modest projects in diverse areas with a couple of projects focussing on the existing firms in the motor industry component supply chain. There was also an investment in a mohair processing factory in Port Elizabeth, but the claimed size of the investment and the projected export sales seemed to be on the optimistic side, given the company's fortunes and production dynamics at the time of the study (Bank et al. 2002, pp. 25–27; Haines 2006).

Nationally and regionally, the NIPP programme – the defence-related aspects in particular – experienced a range of strategic and technical shortcomings in the early and mid-2000s. These included the following:

- Project identification and selection
- Establishing value-added linkages with the regional and local economies in question
- Liaising with local business and government
- Original Equipment Manufacture (OEM) not fulfilling their milestones and obligations timeously
- Ensuring that the interventions had some integrative economic logic – that the projects were not overly diverse and dissipated (Haines 2004, 2007)

There were also found to be substantial hidden costs associated with the offsets, particularly those linked with the SDP. For instance, substantial state investment in regional infrastructure and other resources would be needed for certain of the larger and capital-intensive offsets to work as planned, but were not forthcoming. Yet this kind of cost is not usually factored into official assessments of the civilian defence offset work (Haines 2004, 2006, 2007; Hosking and Haines 2005, pp. 1–23).

In addition, the DTI section responsible for administering the NIP programme and monitoring the IP ventures seemed under-resourced and stretched in the early and mid-2000s. This made hard-nosed negotiation with obligors more difficult. Achieving the necessary vigilance and rigour in the awarding of credits and overall monitoring of projects was also a challenge. Policy lessons were learnt and, by the end of the decade, the DTI had made some progress in tackling a number of these challenges and better coordinating the growing defence and non-defence portfolio of the NIPP projects and ventures (Haines 2004, 2006, 2007).

Contemporary Developments and Issues Regarding the NIPP Programme

In more recent years, as can be seen in the official analyses outlined earlier in the chapter, the NIPP interventions expanded and gained more coherence in focusregarding projects and implementation. There is now an array of projects, but closer independent analyses requires more case study material. Such inquiry is also undercut, to an extent, by the difficulties of extracting relevant data from either official sources or the firms and ventures involved. Nevertheless, certain questions can still be raised about the nature and efficacy of the programme – both in civilian and defence-related aspects.

For one, the geographical spread of projects currently has mostly confirmed the current spread of economic production within South Africa and the dominance of Gauteng and certain secondary metropolitan centres, namely Durban and Cape Town. The poorer and more rural provinces, with the partial exception of the Eastern Cape, have not been allocated much collectively. The situation is represented in Fig. 6.2 and Table 6.1. The SDIs and IDZs – interventions designed i.e. to achieve cost-effective regional and local development – have not received sufficient

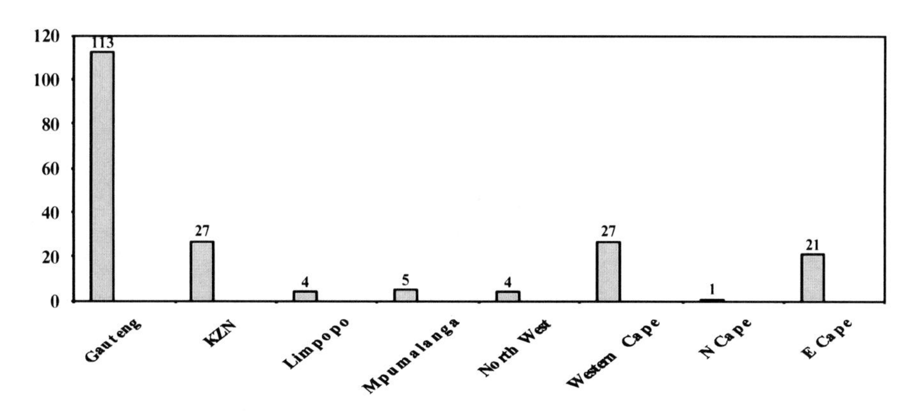

Fig. 6.2 Provincial distribution of national industrial participation projects in South Africa 1997–2008. *Source*: De Risi 2009

Table 6.1 GDP per head by province, 1996–2008

	GDP per head				
	R[a]				
Provinces	1996	2000	2004	2008	Real per head growth (1996–2008) (%)
Eastern Cape	11,188	11,834	12,861	14,883	33.0
Free State	17,925	17,926	19,162	21,976	22.6
Gauteng	33,379	34,779	38,112	44,735	34.0
KwaZulu-Natal	15,647	16,035	17,681	20,753	32.6
Limpopo	10,802	11,477	12,934	14,651	35.6
Mpumalanga	18,568	19,327	20,366	22,286	20.0
North West	19,437	18,102	18,964	21,294	9.6
Northern Cape	20,422	21,843	22,176	23,952	17.3
Western Cate	29,049	29,422	32,608	38,214	31.8
South Africa	19,989	20,714	22,708	26,335	31.8

Source: SA Survey (2008–2009, p. 101)
[a]At constant 2000 prices

projects to achieve any critical mass in regard to (partial) decentralized development (Bank et al. 2002, pp. 25–27; Haines 2004, 2009a; Wellman 2010).

Second, one of the objectives of the NIPP programme, most especially the SDP side, was to contribute to the diversification of South Africa's industrial base. Whether this has been achieved in a meaningful way is arguable. A number of the NIPP projects and investments have been in existing firms and enterprises. There could have been more in the way of greenfields investment, and in emerging high-tech industries. The creation of a venture capital fund for such a purpose with leveraged funding by obligors would have been both strategic and practical. There are, nonetheless, certain new kinds of productive interventions most notably being in the manufacturing of interior parts for Boeing commercial aircraft by Aerosud Aviation, and the incorporation of the company in Boeing's international supply chain (Haines 2009b).

Indeed, in certain respects there has been a loss of manufacturing capacity in certain sectors during the 2000s. The reluctance to use IP interventions to help sustain crucial productive assets regarding South Africa's shipbuilding industry, as discussed earlier, is one example. If one considers the overall decline in productive capacity within South Africa's defence industrial base – discussed later in the chapter – then this argument is reinforced.

Third, systematic research is needed to determine the extent of technology transfer via the NIPP programme. Given that a wide range of activities are included in the rubric of technology transfer in both the DIP and the NIP programme, and the general reluctance of obligors to transfer higher-end propriety and strategic know-how (e.g. Dunne and Haines 2006), one would be surprised if the actual technology transfer was particularly pervasive or profound.

In a 2009 presentation, the DTI listed three examples of technology transfer. The first was Aerosud's relationship with both Boeing and BAE-SAAB where "obligors provided know-how transfer, building up capability on manufacturing and engineering, tooling production and training". This has led to Aerosud now being recognized as an aerospace first-tier global supplier and well integrated within the supplier network. The second example is the Mechatronics Training Center (Thyssen Krupp Marine Systems) partnered with the Higher Education (HE) institutions such as NMMU where the chief objectives were to inculcate integrated mechanical and engineering skills with IT and Informatics – which were central to the value-added manufacturing sector and the South African automotive industry (Msomi and Shelver 2010). While the project is still a work-in-progress, the HE and Further Education (FET) sectors in South Africa are areas where international partnerships are both cost-effective (for the obligor) and provide reasonable value for the recipients (Msomi and Shelver 2010).

The third example is the role of the BAE-SAAB and Agusta in projects linked to the beneficiation of South Africa's rare metals and minerals including gold, platinum and diamonds – most especially in regard to jewellery manufacturing. Both the BAE-SAAB and Agusta invested both funds and technology in establishing these manufacturing facilities and have also provided training (De Risi 2009).

Gwendolyn Wellman's doctoral study which deals primarily with the BAE-SAAB South African Royal Manufacturer's project – a greenfields gold jewellery venture run in conjunction with Harmony Gold (a gold mining enterprise located in Virginia, Free State province) – explores why this flagship initiative failed after 12 months of operation. Such a project, which had backing from the Mintek (a parastatal focusing on mining-related technology and support), the Industrial Development Corporation, and the DTI, had all the elements to be successful. Instead, Wellman found a relatively low level of FDI from the obligor with much funding leveraged locally or supplied at soft rates by the IDC; poor monitoring of the project; and a rather vague process of allocating credits. The relevant DTI official at the time candidly admitted that the NIPP division was under-resourced and that the turnover of personnel meant that there was not always sufficiently detailed understanding of certain projects (Wellman 2010).

By contrast, the continued success of two Western Cape projects – OroAfrica (a gold jewellery beneficiation enterprise) and Silplat-Platinum (a jewellery beneficiation venture) – in which Agusta (OraAfrica) and BAE-SAAB (Silplat) took a stake in as offset projects, offers certain policy insights (Wellmann 2004; Wellman 2010). These two projects demonstrate "that if conceived well with a proper due diligence investigation in place; combined with real foreign direct investment, and a genuine interest by the foreign investor to share international markets (or that there are actually existing markets to share); and provided that the company produces good quality products to international standards; small individual offset projects can be successful" (Wellman 2010, p. 262). The point can also be made, however, that these projects had been going concerns for a significant period before the offset investment.

Agusta also looked to invest in the beneficiation of mohair through its link with Cape Mohair. Yet, the beneficiation of the mohair was achieved in part through sub-contracting to another firm in Plettenberg Bay some 200 km from the factory (Dunne and Haines 2001; Haines 2004). While such investments in the beneficiation of luxury raw materials such as mohair and gold and precious metals are not without merit, there is scope for more substantive sectoral support, and consideration of the issue of regional concentration and specialization in the establishment and extension of the requisite value chains.

On the other hand, it is also important to bear in mind, as Riaan Coetzee of the IDC pointed out in 2001, that the gold and mohair industry projects had a long developmental history in South Africa dating back to the 1980s. Their development has been on the cards for some time, and the question has been raised whether their export promotion has come about as a result of the SDP and NIPP or whether it fitted conveniently in (cited in McEwan 2002).

As Brauer and Dunne argue, if not linked specifically with military industry, most countries focus their offset interventions on technology transfers tied to local industry. South Africa is one of the few countries which "still seem[s] to believe that offsets can result in across-the-board generalized economic development and job creation" (Brauer and Dunne 2010).

Job Creation

Responding to the DTI's 2009 performance review of the NIPP that was recently tabled in the Parliament, Tim Harris, the shadow minister of Trade and Industry, pointed out that the SDP had only generated around 26,000 direct jobs instead of the 65,000 originally forecast. He complained, in addition, that it is "close to impossible to assess the status of the NIPP because the DTI continues to obfuscate in its reporting".[4] Ironically, it is quite likely that the figure of 26,000 jobs is itself on the optimistic side. Moreover, as Dunne and Lamb point out, while a figure of 65,000 jobs "sounds impressive [it] amounts to a cost of R1.6 million per job and is

[4] *Business Day*, 14 September 2010, *Defence Web*, 13 September 2010.

extremely high, nearly 20 times the average cost per job in South Africa's defence industry" (Dunne and Lamb 2004). The 2008 exposés of the SDP by the Sunday Times (of South Africa) and the Mail and Guardian (also South African, but with a link with the UK Guardian) resulted in a government briefing in August of that year. An extract from the official transcript of the meeting, which covers the interaction between an investigative journalist and Alec Irwin, then the Minister of Public Enterprises, and formerly the Minister of Trade and Industry during the late 1990s and early 2000s, is worth quoting at some length:

> [Alec Irwin]: Our estimate is that at this point probably around about 20,000 direct jobs have arisen from the two packages, and probably another 30,000 indirect jobs. And that's fairly close – and we haven't finished the programme yet – to the 66,000 jobs we spoke about in a hearing in [1999]...
>
> Journalist: Minister, just on the subject of the offsets. I think it was in 2005 that I got the Trade and Industry figures for job creation and I phoned up a number of companies involved to check whether they had as many people employed as the documentation supplied to the committee said, and in all of the cases they said, in fact, that they didn't and that those were the numbers of people that they hoped to eventually employ. I can remember one company in particular, called Blackstone Tech, that manufactured carbon fibre motorbike wheels and I think the document said it had thirty or forty employees and it had maybe fifteen. We've subsequently spoken to the Auditor-General's office about the degree of rigour with which they audit those figures and they've told us that what they do, <u>is not an audit, but a review</u>[emphasis added]. We haven't been able to establish exactly what a review means, and then finally, in the case of defence offsets, it's very difficult for us to independently try and verify what's going on there because repeatedly we're told that even the names of the companies involved are confidential. So the crux of my question is, given all the circumstances, how can we really have faith in the numbers that we keep on hearing about the offsets?[5]

While scholarly research on job creation figures is mostly restricted to certain case studies, there is a distinct and continuing tension between official and unofficial estimates. Internationally, such data has been difficult to extract and confirm, with South Africa being no different to the norm. One touchstone, however, is supplied in Ron Matthews's early study of the Saudi Arabian DIP programme wherein he found that around 2,000 jobs were generated rather than the projected 75,000 (Matthews 2002).

In 2003, Christopher Wrigley, on the basis of comparing the costs and job creation scheme of the mooted capital-intensive stainless steel plant at Coega, argued that "that [i]f this were typical, the number of new jobs generated by the arms deal would not be 65,000, but around 12,000" (Campaign Against Arms Trade (CAAT) 2003, p. 16). However, had the Coega steel mill been erected this would have been a new investment. Estimating the job creation function in projects linked with existing enterprises was even more complex. What is interesting is that over time there were more references in the relevant DTI reports about the retention of existing jobs rather than the creation of new ones. In addition, the shift of personnel

[5] Transcript of Strategic Briefing Issued by Government Communications (GCIS), 6 August 2008.

from a production line or process, which is being phased out or temporarily suspended, to a new production line associated with an offset project could be seen as constituting a set of new job creation possibilities.

Further Areas of Concern with Regard to the Implementation of Offsets

Among the areas requiring further research are the nature and extent of the multipliers acceded to in the NIPP exercise, as well as the allocation of credits more generally. For instance, Wrigley cites the case of the BAE-SAAB project aimed at upgrading the Harmony gold refinery, which would produce gold at a slightly lower rate than an existing refinery. However, he points out the companies will claim as offset, not the difference, but the value of the full output at the prevailing price per ounce. "By such means an outlay of $70m will enable BAEs and SAAB to meet their combined obligation of over $2 billion in investments" (Campaign Against Arms Trade (CAAT) 2003, p. 17).

There is a sense that the relevant state agencies have at times been generous in decisions regarding the imposition of penalties for non-delivery or delayed delivery. In addition, it could be argued that if a company has defaulted on its obligations under the NIPP and/or programme sanctions against the obligor in terms of new contracts, it should adhere to across offset-related programmes and the relevant state departments.

Finally, there seems to be a need for more direct linkages between the DIP and the NIPP projects. For instance, procurement of aircraft and associated offsets should open up opportunities for investment in the South African aerospace industry directly, as this industry falls in a grey area in which military and civilian sectors overlap considerably.

The Advent of the Competitive Supplier Development Programme

In early 2007 the Government adopted a National Industrial Policy Framework (NIPF) which aimed at a more focussed approach to industrialization. The implementation of this framework was outlined in an Industrial Policy Action Plan (IPAP) led by the Ministry of Trade and Industries. The first IPAP was mostly concerned with establishing the springboard for future actions. A revised IPAP was formulated for the 2010/2011–2012/2013 financial years. Localization, employment and capital leakages were among the strategic themes – reflecting the structural challenges presented by the South African economy. These challenges included structural imbalances in South Africa's growth path; manufacturing; employment, low relative profitability of manufacturing; the costs and allocation of capital; and the failure to leverage procurement sufficiently (Transnet 2010; Ritchken 2009).

In December 2007, the Department of Public Enterprises (DPE) established a Competitive Supplier Development Programme (CSDP) to focus on improving procurement practices within its ambit. This Programme involves "procuring in such a way as to increase the competitiveness, capacity and capability of local supply bases where there are comparative advantages and potential competitive advantages of local supply".[6]

The CSDP is intended to be of particular benefit to the state-owned enterprises (SOEs). Its anticipated benefits include the following:

- Greater security of supply
- Reduced costs of goods and services supplied to the SOEs by either increased competition or lowered logistics costs
- Establishing the advantages of local supply vs. imports which, in turn, include the following:
  - Decreases forex-related risks or premiums
  - Lower stock level requirements
  - Improved communication
  - Shorter delivery times
  - Improved access to skills
  - Greater responsiveness: ability to customize design and innovate around local requirements

Although the CSDP was channelled through the DPE, there has been a multi-agency approach to the initiative with the Ministry of Trade and Industry, with the Department of Science and Technology also being involved in the process. In addition, the UNIDO has been brought in as partner to help in the conceptualization and implementation of the programme (Phillips 2009). By looking more to the local suppliers and the contingent industrial complexity in the relevant clusters will require a significant decrease in the percentage of imported capital goods. However, as Edwin Ritchkin, chief advisor to the DPE, pointed out, a 25% drop in imports will have a disproportional impact on growth and output, and will necessitate qualitatively new investments to decrease the imported component. Such a restructuring and clustering of relevant industrial activity could not be accommodated within the ambit of the NIPP. The arguments for the DPE's position on this are fivefold. First, the accountability and responsibility for developing suppliers and decreasing imports is preferably located with the procuring organization. By contrast, the NIPP involves a contract between supplier and the DTI, and effectively removes responsibility from the procurer. Second, there is a desire to ensure the quality of localization proposals forms part of the competition between suppliers, as opposed to a post-procurement administrative process between suppliers and the DTI. Third, distinct and measurable outputs of the process of reinforcing the SOE supply chains as strategic infrastructure are required, "rather than indirect NIPP which prevents this from taking place"

[6] DPE Concept Note cited in Transnet (2010), p. 3.

(Ritchken 2009). Fourth, the delivery of industrial obligations should be embedded in the contractual process with associated penalties. This would be preferable to the prevailing NIPP penalty on imported value. Finally, the NIPP has not been central in the relevant industrial strategy and policy interventions of the rapidly industrializing countries; the focus rather was on constructing the ability to procure developmentally (Ritchken 2009).

The period 2008–2010 has seen the shifting of certain existing ESKOM and Transnet projects within the NIPP programme to the CSDP programmes of these SOEs (Van der Walt 2009; Langenhoven 2009). In addition, new programmes are being conceived with explicit CSDP criteria. Not all the DPE enterprises have been shifted to the CSDP platform, the most notable being Denel. The relevant CSDP programmes are still in the early phases in the relevant policy and implementation process, with the UNIDO playing a role in supplier bench-marking, assisting in the formulation of supplier development plans, and in targeting the focus areas identified in supplier development plans. There is growing enthusiasm from relevant industry groupings, most particularly in the area of steel production and construction (e.g. de Beer 2009) The government is currently looking to improve the linkages and coordination between the NIPP and the CSDP programmes,[7] but this will prove conceptually and logistically challenging. We are still awaiting in-house and scholarly analysis of the progress of the programme, but it would seem that the CSDP approach will yield an improved leverage of procurement for industry-related investment (Lambson 2009).

The SDP, the DIP and the South African Defence-Related Industries

The SDP offsets, especially the DIP components, provided the South African defence industry with something of a lifeline, given the internal political ambivalence to it in the early post-apartheid years. But at the same time the SDP undercut any remaining aspirations for South Africa to maintain its own defence industrial base.

Given the more compact nature of the DIP programme, and more specific sectoral focus, the DIP encountered fewer problems in project selection and implementation than the NIPP side – most particularly in the early stages of the two programmes. Nevertheless, as raised earlier in the paper, there were hidden costs to the SDP and DIP interventions.

Defence suppliers, drawing in part from international practice, tended to include the costs of the offsets and even sometimes the penalty clause in the pricing. The equipment purchased was mostly from different countries of origin than the previous equipment. Furthermore, the local industry had emerged to develop new

[7] *CTO*, 26 September 2010.

Table 6.2 The status of the four defence industrial participation (DIP) portfolios on 31 March 2009

Portfolio	Number of running contracts	Number of completed projects	Total obligation (Rm)	Credits passed during the current financial year	Total credits passed to date
SDPs	7	4	15,111	652	13,158
Active (SDA)	13	19	5,550	92	5,289
Police contracts	2	1	118	17	98
Pro-active	25	8		321	902
Total			*20,779*	*1,082*	*19,447*

Source: Armscor Annual Report, 2009

versions of this equipment and supply some of the parts required. In other words the infrastructural investment, the social capital stock and networks were not sufficiently appreciated in the finalization of the decisions regarding the equipment. With more sophisticated and essentially foreign – as opposed to partly home-grown – equipment, the maintenance work and, thus, the revenue for local defence companies and the associated supply chains were significantly reduced (Dunne and Haines 2006; Haines 2009a).

The publicly owned defence production group Denel and certain private companies were drawn further into the international circuits of defence production, through strategic partnerships, partial or full buy-outs and a general restructuring of the local defence industrial base. While this opened up certain markets, it also diminished the relative autonomy of local industry and expertise, and compromised local export initiatives. A key outcome was the effective shift of the locus of procurement from South Africa to an international set of suppliers (Dunne and Haines 2006; Haines 2009a).

The DIP obligations regarding the SDPs, presented, Armcor, admitted in its 2009 Report, "a seemingly insurmountable challenge" (Armscor 2009, p. 11). However, as the constituent DIP programmes were drawing to a close, with four successfully having been completed, there was cause for a measure of satisfaction. Overall, by 2009, 87% of the total obligation had been met, with the remaining 13% not considered a risk as two of the programmes – for the Maritime Helicopter and the Gripen – were only scheduled for completion by 2010 and 2011, respectively (Armscor 2009, p. 11). With regard to the German Frigate Consortium, the platform obligation had been fully discharged, but there were some difficulties with the Combat Suite with R1,605 million discharged of an overall R2,377 million. This was due to the difficulties experienced by MBDA Missile Systems, the subcontractor to Thales Naval France (TNF). In order not to risk losing benefits to the industry, a portion of the TNF obligation (R949 million) was separated from the original TNF DIP obligation, and a new contract was negotiated with the MBDA (Armscor 2009, p. 12). For more details on the status of the DIP portfolios and schemes, see Tables 6.2 and 6.3.

Table 6.3 The DIP status of the strategic defence packages on 31 March 2009

Project	Obligation Rm	Planned performance Rm	Actual performance Rm	Actual vs. planned %	Actual vs. obligation %	Sales (local and exports) Rm	Technology transfer Rm	Investments Rm
Corvettes	2,941	2,460	2,169	88	74	1,623	521	25
Submarines	1,121	1,121	1,239	111	111	867	364	8
Light utility helicopters	1,194	1,194	1,194	100	100	675	487	32
Hawk aircraft	4,252	4,252	4,252	100	100	3,262	973	17
Gripen aircraft	5,050	4,056	4,012	99	79	2,022	1,817	173
Maritime helicopters	553	374	292	78	53	258	31	3
Total	*15,111*	*13,457*	*13,158*	*98*	*87*	*8,707*	*4,193*	*258*

Source: Armscor Annual Report, 2009

The SDP and Private Sector Defence Firms

Within the private sector, the general expectation was that the larger defence firms would be more favoured in the application of the SDP and subsequent procurement exercises, contributing to the shrinkage of the defence sector. However, the larger defence companies have a more mixed experience of the offsets. The DIP contracts did not always materialize or were less substantial (Haines and Wellman 2005; Dunne and Haines 2006). Such contracts also crowded out other export opportunities. In addition, several smaller independent firms, which were not directly linked with the larger private contractors and/or empowerment companies, received no new contracts. Furthermore, there has been a marginalization of certain leading-edge companies such as C2/I2 which did not fit with the defence obligor's value chains. Other concerns of the private sector firms included the following:

- There was a relatively limited transfer of technology
- There was a shrinkage of local opportunities, a leaching of local technology and a loss of intellectual and social capital in the industry
- There was little or no planned domestic conversion
- There was still some DIP work and credits unallocated
- There was seemingly little interest by the DTI in providing state incentives to domestic defence industrial work

Among the beneficiaries of the SDP and the DIP more generally has been ATE (Advanced Technologies and Engineering), the South African subsidiary to and main industrial facility of the French-based ATE group. The ATE's main focus is developing stand-alone weapons systems for the Eurocopter. The production of digital avionic suites for BAE Systems for the Hawk upgrades is among a number of projects in the pipeline (Heitman 2010).

The Reutech defence divisions, and SAAB Grintek, both of which have seen buy-in (with the DIP credits) from European defence groups, have also had a revenue stream from the DIP work. However, given the long-term financial sustainability of these companies, the overall contribution of the DIP interventions in regard to their income appears relatively modest (Haines 2009a, b).

The SDP, DIP and Denel

The effects of the SDP and the DIP more specifically on Denel, the state-owned defence production group, were complex. The new procurement exercise had a positive impact in the short-to-medium term on the company's fortunes. Yet, more recently, the company has experienced tougher times, and been forced to undergo significant restructuring and job losses. In part this is due to the government indecision on the group's institutional future and the degree of state funding to be allocated in future (Heitman 2010).

Overall, Denel's experience of the SDP did not match all their expectations. In the earlier phase at least, the failure of certain obligors to fulfil their targets and to help stimulate output and exports orders of the Denel group was noted (Ferreira and Haines 2005). Even after 5 years, there was still a shortfall between the promised DIP contracts and the contracts actually received, though this has been mostly addressed of late (Haines 2009a, b).

Other concerns during the period 2000–2006 included the following:

- Difficulty in securing multipliers for contractors.
- Certain company group expected projects, but these were either on a smaller scale or did not materialize.
- Offsets in the aerospace sector crowded out other potential projects in aircraft production, as many small parallel activities loaded the production facility.
- Multi-task SDP was probably difficult to handle logistically: single orders over a period of time would have been better (Dunne and Haines 2006).
- There was a feeling that certain obligors for both the DIP and the NIP offsets tend to plan their discharge scheduling on the "hockey stick" principle, with the bulk of the activities shifted as far as possible to the end of the period in question.

There were, however, a number of achievements and developments that the SDP and the DIP processes contributed to within the Denel group during the mid- and late-2000s, although these too were not without contradictions and complexities. Central to these processes were the restructuring of the group, and the transferring of partial equity and control of several divisions to foreign partners. The DIP credits were allocated for both the buy-in and the existing and new contracts. The purchase of equity by the foreign Original Equipment Manufacturers led to the formation of Carl Zeiss Optronics (30% Denel), Rheinmetall Denel Munitions (49% Denel), Turbomeca Africa (49%) and Denel SAAB Aerostructures (DSA) (80% Denel). According to the 2010 Denel Annual Report, "all three of Denel's associated companies (Turbomeca Africa, Carl Zeiss Optronics and Rheinmetall Denel Munition) have generated increased revenue earnings and profits in the past few years" (Denel 2010, p. 146). DSA has experienced a dramatic decline in orders with the SDP programme nearing an end, and become the biggest loss-maker in the group (Denel 2010). The DSA's financial performance remains a major concern, the business posting losses of R328 million (2009: R452 million) (Denel 2010, p. 21). The ill-considered decision in 2006 to order a tranche of A400 military aircraft and participate in the global production of the craft has been costly. The cancellation of the order due in part to delays and escalating costs of the global programme has led to penalties being imposed by the Airbus group. The shift to partial foreign ownership in the Denel group – stimulated in part by the SDP procurement exercise – does not appear to have contributed to the cohesion of the group and may have come with externalities and hidden costs.

Apart from the DIP relating to the SDPs, there has been the IP on other defence procurement – most notably programmes such as Cytoon, GBADS, Hoefyster, Kingfisher and Klooster – which have proved modestly profitable in recent years (Denel 2010; Armscor 2009).

Defence Industrial Procurement and the South African Defence Industrial Base

The relationship of the South African defence industrial base with the SDP offset programme and external offset packages more generally is a study in ambiguity. While the SDP may in the shorter term have served to prolong the life of this defence industry as a number of analysts have observed, it has also attenuated the industry. Denel has experienced a loss of jobs during the 2000s – from around 11,000 to 8,000 employees with further job losses anticipated. In addition, the private sector firms directly linked to the defence work have declined from around 120 in the late 1990s to around 45 today. Cognizance should also be taken of the loss of expertise within both Denel and Armcor during the implementation of the SDP. For instance, in its 2009 annual report, Armscor complained that it continued

> ...to lose key personnel and, as a result, loses critical capabilities. Thus Armscor's ability to provide a quality service to its customers and to meet its Service Level Agreement (SLA) objectives is negatively affected (Armscor 2009, p. 10).

The question also needs to be asked why South Africa saw fit to move away from a strong local emphasis in procurement to an extensive reliance on overseas suppliers. In addition, the shift from established links with French and Italian companies in regard to the fighter and the LIFT aircraft has impacted on the sunk costs, the social capital, and productive and supply networks established around the local beneficiation and indigenization of Mirage and Aeromacchi aircraft. As the Denel CE Ismail Dockrat remarked, in 2009 during a marketing exercise to sell Cheetahs abroad, "Denel Aviation has an accumulation of 40 years of skills, expertise and business know-how in performing maintenance, repair and overhaul (MRO) work on the Mirage type aircraft".[8]

The cost of a military aircraft is around a third of its lifecycle costs, and by changing the origins of the equipment procured, one is effectively changing the logistics philosophy of the recipient airforce. This, in turn, has a range of cost and resource implications, including the retraining of technical personnel. The shift to direct procurement of the UK and Anglo-Swedish aircraft appears to have had an adverse impact on South Africa's aircraft manufacturing and modification facility established during the 1980s and 1990s. While the offset work has provided orders for a section of the local defence aerospace base, a good deal of new work has been in the order of components – tail fins, pylons etc. – rather than larger, more substantive and sustainable aspects of aircraft production and associated technology transfer. There are exceptions, the ATE and Turbomeca operations being cases in point, but overall there has been a marked downturn in fortunes in regard to the production of indigenous military and civil-military aircraft production in South Africa (Haines 2009a).

[8] Cited in *Defence Web*, 8 October 2009.

If one considers the extent of credits associated with the Gripen programme and the developmental costs of the non-NATO base line, mostly reflected in the up-front price, the current lack of orders experienced by the DSA and the subsequent severing of the link with the Mirage indigenization and remanufacturing process becomes more questionable.

A further case in point is the South African-produced Rooivalk attack helicopter. Links with overseas OEMs have failed to benefit Denel and its marketing. The UK Ministry of Defence was considering to include the aircraft in its set of orders for new helicopters in the early 2000s, when active US lobbying and an insistence on US Nato-specific weapons system for the procurement exercise in question saw its omission from a list of candidates. Some years later, the Rooivalk had won a technical evaluation along with the Mangusta helicopter in a Turkish procurement process for an attack helicopter. However, the Denel senior management were not sufficiently insistent in persuading the Eurocopter – which was linked to Denel via Turbomeca – to help underwrite their bid and provide the requisite drive trains for the Rooivalk (Haines 2009a).

Overall, then the nature and extent of technology transfer have probably been overestimated. Indeed, in certain cases there has been a reversal and/or dilution of local technology. A particularly interesting case was that of Cape Town IT and the defence company C2I2 which had advanced combat suite technology and looked to have the inside track to gain the relevant sub-contract to install such systems on the frigates. However, the obligor opted for what seemed to be a less advanced system. The CEO of the company, Richard Young, argued that the then DoD chief of acquisitions had a conflict of interest in the matter, and also subsequently, and successfully, sued the South Africa government (Haines 2007).

The impact of the SDP and other offsets on the South African defence industrial base is thus difficult to cost, and whatever exercise undertaken depends in part on the methodology and the degree to which externalities, hidden costs and the dilution of social capital is taken adequately into account. What the South African government has not fully appreciated is the White Paper on National Defence (RSA Government 1996b). The White Paper on South African Defence-Related Industries (Republic of South Africa 1999), and subsequent emerging policy exercises is the profound inter-relationship of the defence industrial base with the South African industrial economy more generally. In an implicit appeal to the national government, the most recent report of Denel stressed that:

> ... [T]he defence industry is a fertile ground for nurturing engineers, technicians and artisans, with most of these technical professions subsequently contributing to key national projects in transportation, construction and power generation and wider general advanced manufacturing activities. By leveraging off the defence technology base, innovative applications have found good use in rail safety, crime prevention, surveillance, protection of assets, mine safety management, mining drill-bits, commercial brass strips, amongst others. Another key contribution is in the form of high value-added exports that improve not only the country's foreign reserves, but also its balance of payments (Denel 2010).

Concluding Remarks and Policy Recommendations

This chapter has provided a select and impressionistic analysis of the NIPP, with a particular emphasis on the offset projects related to defence procurement. It has also considered the DIP programme and the impact of the SDP on the defence industrial base.

In terms of the official record, the NIPP appears to have achieved solid success in terms of the array of projects, and, implicitly, a reinforcement of the key sectors within the national economy, as well as signs of the diversification of the industrial economy. Job creation appears to have, more or less, born out of earlier official estimations, although the distinction between direct and indirect jobs was not made during the rolling out of the NIPP programme (both SDP and non-defence aspects) during the period 1999–2001. Policy lessons have been learned, and modifications have been implemented, and, overall, a greater confidence and coherence can be detected in the nature, range of projects and even the administration thereof. Closer analysis of the NIPP and the related DIP interventions would require a host of detailed case studies. In addition, a detailed development history would provide useful insights for future reflexive policy and implementation methodology.

What bedevilled the SDP part of the NIPP programme, most particularly in its earlier stages, was the need to get a variety of projects up and running in a relatively short time. Meshing South African government expectations with obligor capacity and business instincts took time. There was an over-optimism in certain cases. The growing experience has led to a somewhat more cautious approach. While it is difficult to generalize on project selection, especially with hindsight, there are some pointers. Some of the larger, ambitious and/or capital-intensive projects proved problematic, such as the steel mill's and associated metals' beneficiation originally proposed for the Coega IDZ. Certain greenfields projects, without a substantive managerial culture in place, were less likely to succeed, such as the SARM gold beneficiation project in Virginia. Also, obligors taking over the reins of ventures with a troublesome development history, such as the Magwa Tea Estate venture, were also putting themselves at risk.

The current nature and spatial deployment of the NIP and the DIP offsets tend to confirm, rather than challenge, the existing spatial patterns and inequalities in the South African economy. The employment creation possibilities, offered by the offset projects, particularly for more peripheral regions, are limited. There is a strong emphasis on capital-intensive industries, and a tendency to opt for brownfields rather than greenfields projects, with astute piggybacking and financial leveraging by obligors in a number of instances. The spatial targeting of projects has been carried out in a relatively informal basis, with a reasonable degree of discretionary power left for the obligors. As the major recipient of the DIP and the NIPP offsets, and a key beneficiary from other Industrial Participation programmes, the hegemony of the Gauteng region in the national economy seems set to continue.

The DIP and the NIPP schemes need to be planned and evaluated within a wider sub-national context that recognizes the spatial inequalities in the South African economy. More attention should be paid to the question of sustaining strategic industry capacity. Policy makers need to address the question of utilizing natural resources in a sustainable and equitable fashion, and explore the potential synergies between such resources and broader-based industrial and high technology development.

This also requires that shortcomings in institutional capacity within state agencies involved in industrial development be addressed, and that more creative partnerships with non-state agencies be considered. In general the relevant policy makers must, thus, be more reflexive in their evaluation of and response to the implementation of the NIPP and the DIP projects, and to take note of unanticipated outcomes and issues of transparency.

A further finding is that there are substantial hidden costs associated with certain offsets – especially the large and more ambitious and capital-intensive ones. This is particularly so for those offset projects originally selected for the Coega IDZ. The offset projects can and do require additional state investment in infrastructure and other resources to facilitate the project or investment.

The current set of NIPP interventions and projects could do more in the way of strengthening of linkages with secondary industry and with the economy at large. What is somewhat surprising, for example, is that the possibility for utilizing increased and more diverse steel and aluminium products in industries such as shipbuilding, which have demonstrated a capacity for niched export production, have not yet been systematically explored. However, there are signs that this concern may be addressed in time by the CSDP initiative.

While the DIP offsets have been more focussed and compact than those in the NIPP stable, there are a range of structural shortcomings. And similarly to the NIPP offsets, there are significant hidden costs ranging from the up-front and maintenance costs of the equipment through to the differing agendas of local defence actors and international obligors.

Focussing on offsets in order to utilize international inputs to diversify industry has diverted some attention away in part to more indigenous means of achieving this aim. The government has tended to take too negative a view of the possibilities of qualified import substitution, and of ensuring the underpinnings of strategic industrial sectors and sub-sectors. However, the IPAP process and the inception of the CSDP suggest that the tide may be turning in this regard. One is not advocating the retention of obsolete industries, but rather to stress the importance of supporting strategic albeit vulnerable industries, and of tolerating a degree of market failure especially in secondary centres and peripheral regions of South Africa. Given the massive loss of manufacturing jobs in the South African economy in recent years, this path is worth pursuing. This would help more to address the problems of unemployment and globalization, For instance, the withdrawal of subsidies to industries in centres, such as East London in the early 1990s, has not been ameliorated, and has contributed to sectoral industrial decline and increased unemployment (Bank et al. 2002) in the Eastern Cape, one of South Africa's poorest provinces. It could be argued that more use should have been made of certain of the offset investments in helping to shore up industrial capacity in selected strategic industries and help stem job losses in the selected sectors such as textiles and clothing, targeted back in the early 2000s by the 2002 integrated manufacturing strategy (IMS) DTI (2002c).

The question of job creation and the related issues around credit and multiplier allocation to obligors are all areas where current methodological approaches do not appear to be sufficiently consistent and reliable. The job creation figures seem far less substantial when one extrapolates from a series of case studies on the working and failed projects. More emphasis should be put on independent audits of the NIPP and the DIP approach, through possibly the creation of a multi-actor offset accreditation and advisory body. In terms of technology transfer, in both the NIPP and the DIP interventions the transfers seem less substantive and more problematic in practice than in theory.

Despite the fact that the SDP has seen an overlap between defence and civil procurement, there is a tendency to view South Africa's defence-related industries as relatively marginal to the industrial trajectory of twenty-first century South Africa. The failure to grasp the significant interpenetration between the military and civilian industry, and a seeming complacency about the continued erosion of the defence industrial base, has distinct knock-on effects for high-tech industrial growth and capacity prospects.

The CSDP initiative is an acknowledgement of the shortcomings of relying too heavily on internationally oriented procurement to leverage local economic growth and diversification, and technological enhancement. However, there are still significant logistical and institutional impediments to the articulation of more coherent and comprehensive approaches to sustainable industrial and economic production and development.

While the continued enthusiasm for supply and value chains within the government is hopefully informing a move to a more focussed set of interventions in the industrial economy, it would be inadvisable to downplay the need for considered targeting and more balanced and sustainable inward development. In considering the possibilities for local and regional economic and industrial development in South Africa, new institutionalist studies of regional economic development should be taken into consideration. This approach challenges the policy makers to shift from firm-centred and state-driven modes of promoting regional and local economic development policy to more horizontal, multi-agency, and decentralized forms of economic development planning. More attention still need to be accorded to the construction of appropriate institutional structures both private and public, the deployment of creative partnerships, and the enhancement of social and human capital.

While there are signs that the South African government is looking to decrease somewhat its reliance on foreign procurement, interventions such as the NIPP and the DIP schemes look set to continue. There is scope, however, for more cost-effective and creative approaches to offset projects in which the social capital and networks a large international company can deploy should be taken more systematically into account (Lambson 2009). The area of higher education partnerships is one example of this. In addition, the possibility of setting up one or more venture capital funds (using funding and leverage of the obligors) to focus on areas, such as high-tech emerging business capable of accessing export markets, and green energy projects, would be worthwhile.

In conclusion, the above discussion and assessment of the NIPP reveal complexities and contradictions in differing interpretations and recording of substantive industrial participation venture. This complexity should be seen a conceptual challenge for both protagonists and critics of offset ventures in Sou Africa and internationally. They will need to deploy more contextually relevant ar reflexive cost–benefit analyses and find innovative ways to secure more reliabl project-by-project data.

References

Armscor (2009) Annual report 2009. DoD/Armscor, Pretoria

Balakrishnan K, Matthews R (2009) The role of offsets in Malaysian defence industrialization. Defence Peace Econ 20(4):341–358

Bank L, Haines RJ, Hosking S (2002) The defence offsets and economic development in the Eastern Cape. Paper presented to ECAAR/NRF international conference on defence offsets, Graduate School of Business, UCT, Cape Town, Sept 2002, pp 25–27

Batchelor P, Dunne P (2000) Industrial participation, investment and growth: the case of South Africa's defence related industry. Working paper. Centre for Conflict Resolution and Middlesex University Business School, Cape Town

Bond P (2002) Unsustainable South Africa. Merlin, London

Brauer J, Dunne JP (eds) (2004) Arms trade and economic development: theory, policy, and cases in arms trade offsets. Routledge, London

Brauer J, Dunne JP (2010) Arms trade offsets: what do we know? In: Coyne C (ed) Handbook on the political economy of war. Edward Elgar, Cheltenham

Campaign Against Arms Trade (CAAT) (2003) The South African arms deal: a case study in the arms trade. CAAT, London

De Beer K (2009) Southern African institute of steel construction: structural steel in CSDP. Paper presented at the first international conference on offsets, countertrade, and industrial participation, Port Elizabeth, 12–13 Nov 2009

De Risi T (2009) The national industrial participation programme. Paper presented at the first international conference on offsets, countertrade, and industrial participation, Port Elizabeth, 12–13 Nov 2009

Denel (2010) Annual report 2010. Denel, Pretoria

DTI (Department of Trade and Industry) (1997) National industrial participation policy for South Africa. Government Printer, Pretoria

DTI (Department of Trade and Industry) (2002a) Report on the industrial participation programme. Final draft, Mimeo, March 2002

DTI (Department of Trade and Industry) (2002b) Leveraging growth in the South African economy: the national industrial participation programme in review. Government Printer, Pretoria

DTI (Department of Trade and Industry) (2002c) Accelerating growth and development: the contribution of the integrated manufacturing strategy. Government Printer, Pretoria

DTI (Department of Trade and Industry) (2003) Report on projects of the national industrial participation programme. Government Printer, Pretoria

Dunne JP, Haines RJ (2001) Defence procurement and regional industrial development in South Africa: a case study of the eastern cape. Middlesex University Business School, Discussion paper series, no. 98

Dunne JP, Haines RJ (2006) Transformation or stagnation? The South African Defence Industry in transformation. Defence Stud 6(2):169–188

Dunne JP, Lamb G (2004) Defence industrial participation: the South African experience. In: Brauer J, Dunne JP (eds) Arms trade and economic development: theory, policy, and cases in arms trade offsets. Routledge, London, pp 284–298

Ferreira R, Haines RJ (2005) Industrial Participation and Defence Offsets in Gauteng. Paper presented to Annual Congress of South African Sociological Association, University of the North. July.

Haines RJ (2004) Defence offsets and regional development in South Africa. In: Brauer J, Dunne JP (eds) Arms trade and economic development: theory, policy, and cases in arms trade offsets. Routledge, London

Haines RJ (2005) The politics of the strategic defence procurement programme in South Africa: snapshots from Kwa-Zulu Natal. Africanus 35(1):59–73

Haines RJ (2006) Defence offsets and ındustrial participation in South Africa: current initiatives and the path for the future. Published Proceedings, SMI conference on ındustrial offsets and industrial cooperation, The Hatton, London, 22–23 March 2006, SMI, London (CD)

Haines RJ (2007) The art of the deal: a re-evaluation of the South African defence offset programme. Published Proceedings, SMI international conference on ındustrial offsets and ındustrial cooperation, The Hatton, London, 21–22 March 2007, SMI, London (CD)

Haines RJ (2009a) In defence of the realm? Towards a deconstruction of the South African arms deal. Paper presented at the NMMU arts faculty seminar series, 4 Aug 2009

Haines RJ (2009b) Snapshots of South Africa's national industrial participation programme. Paper presented at the first international conference on offsets, countertrade, and industrial participation, Port Elizabeth, 12–13 Nov 2009

Haines RJ, Wellman G (2005) Value chains and institutional imperatives in regional ındustrial development: a consideration of the implementation and impact of defence offsets in the Western Cape. Africanus 35(1):25–34

Heitman H (2010) South Africa's defence industry: turning the corner? Armada Into, 1 Aug 2010. http://www.faqs.org/periodicals/201008/2132314051.html. Accessed 21 Sept 2010

Hosking S, Haines RJ (2005) A bridge too far? The South Africa defence offsets and the Coega IDZ. Soc Transit 36(1):1–23

Hurst A (2010) Complexity and the idea of human development. S Afr J Philos 29(3):310–328

Lambson R (2009) South Africa's changing offset environment – a perspective for multi-nationals. Paper presented at the first international conference on offsets, countertrade, and industrial participation, Port Elizabeth, 12–13 Nov 2009

Langenhoven H (2009) Eskom procurement and supply chain management: Eskom supplier development program. Paper presented at the first international conference on offsets, countertrade, and industrial participation, Port Elizabeth, 12–13 Nov 2009

Matthews R (2002) Saudi Arabia: defense offsets and development. In: Brauer J, Dunne JP (eds) The arms industry in developing nations: history and post-cold war assessment. Palgrave, London

Matthews R, Williams R (2000) Technology transfer: examining Britain's defence ındustrial participation policy. RUSI J 145(2):26–31

McEwan G (2002) Defence offsets and the aerospace industry. Paper presented to ECAAR/NRF international conference on defence offsets, Graduate School of Business, UCT, Cape Town, 25–27 Sept 2002

Monitor Company (2000) Durban at the crossroads. Confidential report prepared for the Durban Unicom

Msomi S, Shelver A (2010) Offsets and higher education. Perspectives (2):20–33

Phillips S (2009) UNIDO SPX Programme: supplier benchmarking and supplier development programme. Paper presented at the first international conference on offsets, countertrade, and industrial participation, Port Elizabeth, 12–13 Nov 2009

Republic of South Africa (1999) White paper on South African defence related industries. Pretoria, Government Printer

Ritchken E (2009) Department of Public Enterprises: the competitive supplier development program 7. Paper presented at the first international conference on offsets, countertrade, and industrial participation, Port Elizabeth, 12–13 Nov 2009

RSA Government (1996a) Growth, employment and redistribution: a macroeconomic strategy. Government Printer, Pretoria

RSA Government (1996b) Defence in a democracy, White paper on national defence for the Republic of South Africa. Government Printer, Pretoria

SA Survey (2008/2009) "The Economy", pp. 85–190. South African institute of race relations. Downlaoded from: http://www.sairr.org.za/services/publications/south-africa-survey/south-africa-survey-online-2008-2009/the-economy-1. Accessed on March 2011

Transnet (2010) Transnet supplier development plan. Brochure, April 2010. Downloaded from: http://www.transnet.net/BusinessWithUs/PolicyDocuments/SDP_Brochure_April_2010.pdf. Accessed on: January 2011

Van der Walt F (2009) Transnet competitive supplier development program (2 years on). Paper presented at the first international conference on offsets, countertrade, and industrial participation, Port Elizabeth, 12–13 Nov 2009

Wellman G (2010) An evaluation of the BAE/SAAB South African royal manufacturing project in Virginia, Free State Province: a case study of the implementation of the South African defence offsets. D. Phil., Nelson Mandela Metropolitan University

Wellmann G (2004) An evaluation of the Augusta Westland Filk Gold Chains in South Africa national industrial participation project. MA Thesis, Nelson Mandela Metropolitan University

Chapter 7
Learning-Based Technology Transfer Policies and Late Development: The South Korea Experience

Murad Tiryakioğlu

> *... The third country in the world to introduce and export 256 k memory chips, after Japan and USA was not an OECD or a COMECON country, but South Korea...*
>
> (Freeman, 1989:86)

Introduction

It is an accepted phenomenon by almost all economic schools of thought that appeared after the industrial revolution that knowledge and technology, which have grown to be important benchmarks, have together played a key role in the development process. With the "*creative destruction*" theory based on Marx's theory of surplus value, Schumpeter[1] emphasized that technology included the methods to increase national prosperity and argued that technological innovations were the leading factors contributing to competitive advantage in the capitalist economies. This theory revealed that certain amount of technology was necessary to help organize and ensure the availability of the factors of production in such a way as to ensure growth, secure economic development, and increase national welfare. However, it is indicated that reaching such a level of technology and obtaining access to it constitute problems on their own. According to the World Bank,[2] the poverty that the developing countries suffer is from lack of knowledge, which in turn restricts development.

[1] Schumpeter (1942).

[2] World Bank (1999), p. 19.

M. Tiryakioğlu (✉)
Afyon Kocatepe University, Afyonkarahisar, Turkey
e-mail: tiryakioglu@aku.edu.tr

M.A. Yülek and T.K. Taylor (eds.), *Designing Public Procurement Policy in Developing Countries*, DOI 10.1007/978-1-4614-1442-1_7,

Defined as facts obtained through learning, research, observation, and knowledge is the source of technology and innovation that are produced, processed, used, and spread by the human beings. In other words, knowledge that constitutes the source of technology is defined as the formation methods about an industry branch, the used supplies, and tools. Furthermore, technology is about the used supplies and tools so as to reach the expected results and to use the prepared plans and defines the technology as the application of knowledge.[3] Also defined as the *knowledge of society on production*, the growth of technology is the increase in such knowledge.

The dominant problem of the countries on the development path is the acquisition and production of knowledge. What makes this problem deeper and drags the mentioned countries into a dilemma is that their technological absorptive capacity cannot be enhanced sufficiently enough on account of the poverty in the country. In this context, the scarcity of human capital, organizational difficulties, and lack of social motivation can be counted as the reasons that bring the stability of a knowledge-based development to a standstill. These problems lead, in turn, to a series of defects from the selection of the correct technology to the assimilation and absorption towards production. Finally, the transferred technology cannot be transformed beyond the usage purpose. The successful examples that were able to break this vicious circle and to ensure learning through technology transfer can reach the level of competitiveness with the industrialized countries by using the advantages of "late-comers."

This study is based on learning-based technology transfer policies, and it assesses the technology production by the late-comers, as well as their late-development experiences, through the example of South Korea, which contains lessons to be learned by the other late-comers. The chapter consists of learning-based technology transfer policies and late development, and the related experiences of South Korea as a successful late late-comer.

Learning-Based Technology Acquisition and Late Development

In the classic work of Gerschenkron, published with the title of "*Economic Backwardness in Historical Perspective*" in 1962, late development was brought forward, and it was asserted that late-developing countries such as Germany and Russia needed[4] a strong state intervention in order to catch up with the early developed countries such as England and Switzerland. Gerschenkron (1962) names Germany and Russia in the European continent (and Japan in the Far East) as the late-developing countries that are leading the second-generation industrial revolution and trying to catch up with England that has pioneered in the first

[3] Canberra Manual (1995), p. 16.

[4] Weiss and Hobson (1999), p. 115.

industrial revolution. Referring to this, Vogel[5] names the Asian Tigers – which consisted of South Korea, Taiwan, Hong Kong, and Singapore – as late late-developing countries.

The human and physical poverty of the developing countries are reflected negatively in the production, processing, and commercializing of the knowledge, and it is thus necessary that technology be acquired through outsourcing. Another factor that makes bringing in external technological development necessary is that the potential to produce knowledge is restricted by the brain and knowledge drain caused by the mentioned poverty and deficiencies in these countries. One major problem in these aforementioned countries is that there is a lack of coordination between the side that produces theoretical knowledge (such as the universities that form the technological innovation sources, research and development foundations, research and development branches of the private companies, research institutes, public institutions working on research and development) and the side that turns the theoretical knowledge into economic values in industry. Furthermore, the technological development process was not supported effectively with the help of a National Innovation System.[6] All of these factors will lead to setbacks and will bring the countries (and/or the firms) face to face with the need to undertake technology transfers through purchase and copying.

Technology transfer is usually regarded as an operating system from developed and pioneering countries (or firms) in technology production. The basic reason for this understanding is that the developing countries that do not possess the potential to produce the needed technology can only get access to it – legally or illegally – through transfers. However, technology transfer is not a single one-shot activity, but it is an on-going process. Furthermore, it is not only the case for the less developed countries which were unable to produce technology or the developing countries, but also for the developed countries. It was defined[7] as "*importing the needed technology from the technology producing countries so as to increase productivity and financial growth and development.*" Thus, a remarkable quantity of technology trade and transfer is realized among the developed countries as well.[8] Freeman and Soete highlight the stability of the expectation that basically almost all the technologies will be born in developed countries, but they maintain that new technology could be spread more quickly in a (probably less developed) country that does not know the old technology, has not invested in this technology, and has not produced new technology.[9]

The technology transfer is not only the case for the less developed that was unable to produce technology or developing countries but also for the developed

[5] Vogel (1991), p. 5.

[6] Freeman (1987) and Freeman (1988), Lundvall (1988) and Lundvall (1992), Nelson (1988) and Nelson (1993), Nelson and Rosenberg (1993).

[7] Tiryakioğlu (2006), p. 52.

[8] Pavitt (2001).

[9] Freeman and Soete (1997).

countries. It was defined as[10] "*importing the needed technology from the technology producing countries so as to increase productivity and financial growth and development.*" Technology transfer is performed through three channels: (1) trade in goods, (2) foreign direct investment (FDI) and licensing, and (3) labor turnover and movement of people.[11] According to the World Bank[12]:

> ... Over the past 15 years, the main international channels through which technology is transferred have increased. Developing countries' imports of high-tech goods and of capital goods have risen relative to GDP, and their share in global high-tech export markets has increased. Inflows of FDI have increased six-fold relative to developing countries' output, and opportunities to purchase technology have risen along with FDI outflows...

According to Radosevic,[13] there are numerous criteria which can be used to classify technology transfer, like vertical and horizontal, formal and informal, active and passive role of foreigners, embodied and disembodied, degree of packaging, direct or indirect; institutional form that can illuminate the different aspects of the transfer process (Fig. 7.1). The realization of the horizontal transfer methods has a decisive role in the promotion of the inward transfer of foreign technologies and in the development of absorptive capability to digest, assimilate, and improve upon the transferred technologies. Teece[14] describes the difference between the technology transfer methods classified as horizontal and vertical, whereby horizontal transfer refers to the transfer of technical information from one project to another. It can be distinguished from vertical transfer, which refers to the transfer of technical information within the various stages of a particular innovation process, e.g., from the basic research stage to the applied research stage.

Setting off with this explanation, it could be said that actualizing the technology transfer by vertical method[15] (licensing, know-how, joint venture, purchasing, franchising, ready-to-use facilities, consulting, production partnership, foreign expert employment, and so on) or horizontal method (R&D projects, university-industry cooperation, R&D Institutes, project collaboration, clusters, network structures, intensive interactive systems) is a significant factor that determines and affects the skill to have the technology in the developing countries. In other words, performing the transfer by vertical methods means having no domination and development skill on the transferred technology, and dependence on external sources, whereas performing the transfer by horizontal methods means access to technology, appropriate selection, procuring, assimilation, development, and in the further stages,

---

[10] Tiryakioğlu (2006), p. 52.

[11] Hoekman et al. (2005).

[12] World Bank (2008), p. 150.

[13] Radosevic (1999), pp. 19–20.

[14] Teece (1977), p. 243.

[15] For a study that analyzes vertical technology transfer's effects on industrial development in less developed countries see Pack and Saggi (2001).

it signifies that it is possible to produce technology.[16] As the horizontal transfer methods will make it possible to learn the transferred technology, it will turn this process into an opportunity rather than a threat. Because the knowledge transfer process will provide that technology, production will supply learning from this technology transfer and will reduce dependence on the external sources on the related areas.

The grounds for this technology transfer vary for different parties. The reasons that force the developing countries to transfer technology can be recited as "not possessing adequate capital accumulation and work power," "earning international competition power against developed countries," and the ideological effect that "the use of techniques invented by the developed countries is advantageous."[17] In addition to this, it is also possible to talk about some causes like modernization of the current facilities and ensuring profitability increase. The basic purpose of this learning-based technology transfer is the development of the national technological abilities. The major arguments for the export of the technology transfer can be given as covering their costs by selling the developed technology and gaining profit, selling the technologies so as to extend the sale of the produced investment product or semi-finished products, selling the old technologies used by the companies and gaining profit since they had started to use the new technologies, selling technology (license, know-how, and patent) that ensures the sales of semifinished products to the markets where it becomes hard to sell goods, and gaining additional profit by the sale of widespread technologies.[18] To sum up, the parties that export the technology achieve *profitability;* the parties that import technology achieve *development and competition power.*

For the late-coming country economies, which do not own sources to produce technology and can only own technology by transfer, to form a knowledge-based development model and to maintain this depend on the formation of a learning-based technology transfer policy. Such a policy, which will help develop a skill capability that will assimilate and absorb the acquired technology, will enable the late-comers to catch up with the developed countries.

The first step (Fig. 7.1) of the learning-based technology transfer policy consists of the selection phase. Here, the determination of the technology to be transferred is actualized depending on the transferring country's economic and social structure, technological capability and accumulation, and the potential of human capital. The basic criteria depend on the effective provision of the first step not only for the formation of the policy, but also for the provision of a sustainable success.

The second step is based on learning. In order to form a learning-based technology transfer policy, the most fundamental source is human capital accumulation. The economies that are unable to allocate a share of national expenditure for education

[16] Kiper (2004).

[17] Hamitoğulları (1974).

[18] Işık (1981).

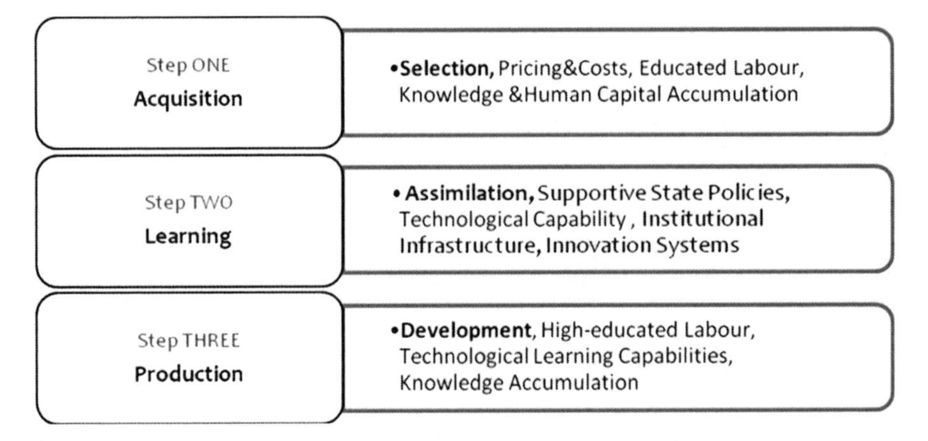

Fig. 7.1 Learning-based technology transfer process and basic factors, Source: Figure 1 was developed by the author

and health at low levels of national income eventually experience human capital poverty,[19] and this result limits the directly acquired knowledge from learning and hinders new knowledge production. At this phase, the state becomes a major actor in both the provision of human capital accumulation and the development of the technological capabilities along with the supporting R&D activities that are to be used effectively.

It is clear that the state's role (together with that of the institutions that form the innovation system) in providing learning from the technology transfer in the long run would have to be to invest in the human capital as the principal factor of technological development by increasing the spending on education and health spending as well as the R&D expenditure equally. In addition, it needs to provide for human capital accumulation and provide technological capability accumulation that will enable the adaptation of the transferred technology and its assimilation. Briefly, the state's pioneering role is characteristic and significant in the provision of general conditions which support the sustainable technological learning and development during the formation and maintenance of the innovation system.[20]

The state grants lying on the basis of a learning-based policy carry an important role in the late-development and late-industrialization process. For almost all the late-industrialization examples, the critical factor in this process is state-supported industrialization, as well as the forming of a special entrepreneur class under the protection of state.[21] F. List, who laid the intellectual grounds of protectionism, argued that in order for a late-developing country like Germany to reach the technological and economic

[19] Tiryakioğlu (2008).

[20] Freeman (1987) and Freeman (1988), Lundvall (1988) and Lundvall (1992), Nelson (1988) and Nelson (1993), Nelson and Rosenberg (1993).

[21] Öniş and Şenses (2009), p. 708.

superiority in England, the country needs to protect its newly established industry with customs barriers.[22] Although List is known and called for the protectionism thesis, free trade philosophy lied under this thought, and List's ideas on catching up with the technology, in fact, constituted a national technology strategy that is set forth, which stated the intellectual grounds of the National Innovation System.

The third and the last steps suggest that the new knowledge and the technology could be producible. In this phase, the trained work force, such as R&D staff and blue collar experts, plays a distinguishing role. This phase in which the organizational constitution helps support this process is the phase that the technological learning efforts show some results. Knowledge production will enable new production design and innovations. In short, the final phase of the learning-based technology transfer points out that the countries could produce their own technologies by themselves.

Guarding against the technology transfer to create a risk for the late late-comers towards technology poverty dilemma and to provide late development by enabling them with the chance to catch up with the industrialized countries can only be achieved through the formation of a learning-based transfer policy and its application. Gerschenkron's thesis, which argues that late-comers may rapidly be industrialized as they have the capability of escaping from some risk and costs pertaining to research and development activities by using the industrialized countries' technological accumulation, displays quite an optimistic tone in the light of the experiences of the countries that gained their independence especially after the World War II.[23] Technological deficiency theory, in which Posner[24] expresses the length of the imitation period, forms the basis of the catch-up problem. Freeman[25] bases his pessimistic reasons on practical reasons that late-comers, which are fed by the accumulation of the pioneering countries in the science and technology area, will face in actualizing the investment and development programs.

Although there are some strong reasons to be pessimistic, the Eastern Asia includes some successful models on late development and catch-up, and this necessarily requires that we evaluate Gerschenkron's thesis one more time. According to Chang and Grabel[26]

> There are five unique characteristics of East Asian countries that were pivotal to the success of this model: *First*, the East Asian countries share a common Confucian culture.... *Second*, East Asian countries are far more ethnically homogeneous than most other developing countries.... *Third*, East Asian countries are blessed with poor resource endowments, and were therefore able to avoid what some call the 'resource curse'.... *Fourth*, East Asian countries benefited in vital ways from (the) Japanese colonialism (that) left behind a strong industrial base, an educated population, and an advanced infrastructure. *Last* not but least, the East Asian model benefited from propitious external circumstances....

[22] List (1841).

[23] Pamukçu (2001), p. 77.

[24] Posner (1961).

[25] Freeman (1989), p. 85.

[26] Chang and Grabel (2004), p. 39.

These fundamental factors all over the Eastern Asia ensured the Asian Tigers, consisting of South Korea, Taiwan, Hong Kong, and Singapore, as late late-developers to appear in the international economic arena. After the Korean War, South Korea has shown great development stability and managed to become different from the other countries in the region. South Korea, which suffered from technological weakness and was almost completely dependent on the foreign technology earlier, was now able to show a success in technology production at a level that can compete against Japan and the United States. The chief factor that lies under the success story of South Korea is that it has learnt from the technology transfer.

South Korea as a Late Late-comer

South Korea, which has some geographical disadvantages such as typhoon, tsunami, and earthquake, had been devastated by the Korean War between 1950 and 1953. When the war ended in 1953, out of the 20 million people in South Korea about a forth were refugees with no homes, almost no assets and, like Chinese mainlanders who came to Taiwan, with little hope of finding secure employment.[27] After such great destruction, the Korean economy that was in physical and human capital poverty followed a development strategy that was almost entirely dependent on foreign resources, and she shaped her development policy on learning-based technology acquisition. Korea's S&T policy that was applied until the early 1980s spent serious efforts towards developing indigenous R&D, which was geared to facilitating learning from foreign technologies, while at the same time developing domestic S&T infrastructure.[28] Efforts in technological capability building and development eventually made South Korea an arbiter in memory chips, mobile phones, and Liquid Crystal Display (LCD) industries. The competitive powers of Japan as representing the first generation of late-comers and of South Korea, in the second generation, draw the following comments from Freeman and Freeman and Soete.[29]

> ... The third country in the world to introduce and export 256 k memory chips, after Japan and USA was not an OECD or a COMECON country, but South Korea....[30]

> ...the enormous success of Japanese, and later South Korean, companies, in catching-up with American semi-conductor companies in the design and manufacture of memory chips, was partly due to this strategy and partly to the recruitment of returning graduate students from American universities and former employees of American companies...[31]

[27] Vogel (1991), p. 43.

[28] Suh (2009), p. 28.

[29] Freeman (1986) and Freeman and Soete (1997).

[30] Freeman (1986), p. 86.

[31] Freeman and Soete (1997), p. 180.

Technology acquisition sources concentrated more upon informal channels than formal ones in Korea's technological learning process. In this process, according to Suh,[32] who argues that the Korean approach to technology acquisition resulted in both positive and negative effects:

> ... On the positive side, this policy enabled Korea to acquire technologies at lower costs, and precluded the constraints often imposed by multinationals on local firms' efforts to develop their own capability. The approach was effective in maintaining independence from the dominance of multinationals. Negative effect is that Korea had to give up an important access to new technologies that might have been available through direct equity links with foreign firms. By restricting FDI, Korea failed to set global standards in domestic business operation. Much worse, large-scale foreign loans that had been brought to finance the massive importation of capital goods, plants and FL contributed to the financial crisis in 1997...

Suh implies that R&D was one of the most damaged victims of the financial crisis[33] which is caused by large-scale foreign loans in 1997, but South Korea could get over the crisis years in a relatively strong manner in the late-industrialization path, yet as can be seen in the quotation below:

> ... If the crisis had continued several more years, the Korean innovation system would have collapsed. Fortunately, however, Korea recovered from the crisis in a relatively short period of time: it took only two years for the industrial R&D to recover and rise over the level prior to the financial crisis...[34]

> ... Asian financial crisis of 1997 struck a serious blow to Korean Innovation System. Private businesses responded to the crisis by cutting R&D investments in a massive way. In the face of declining R&D investments in the private sectors, however, the government increased R&D spending to 5% of its budget, focusing on the development of IT and IT industries. In 2003, government's share in the gross R&D expenditures rose to almost 26% from 20% before the crisis. During this period, IT sectors played key roles in innovation in Korea, leading Korea's recovery from the economic crisis as well as Korea's move toward a knowledge-based economy...[35]

South Korea's development experience, which lasted nearly 40 years, shows that both social and economic factors contributed to the transformation of the country from the poorest war economy to a knowledge economy.[36] However, it will not be wrong to say that industrial transformation was started and continued by two main factors: development stability and communal motivation. Human and knowledge-based policies which were developed with the help of these social factors had played a decisive role in Korea's industrial development. According to Kim,[37]

---

[32] Suh (2009), p. 32.

[33] Suh (2009).

[34] Suh (2009), pp. 34–35.

[35] Suh (2009), p. 29.

[36] For a comprehensive study that analyzes Korean experience and also Asian Tigers' industrial policies see Yülek (1998).

[37] Kim (2005), p. 313.

Korea's industrial development policy can be divided broadly into four phases: "*export drive and development of skilled human resources in the 1960s, heavy and chemical industry (HCI) promotion and building of local technological capacity in the 1970s, trade liberalization and technology-oriented industrial policy in the 1980s, and globalization and information technology industry promotion in the 1990s.*" When South Korea's industrial transformation is monitored in this respect, it is seen that 1960s and 1970s were the learning years based on the assembly process for standard and simple goods. 1980s, as the learning years about design and product innovation skills, indicate a significant transformation. Another important point about the 1980s was the decline in the share of public R&D trend and the fact that the share of the private sector R&D had started to surpass the share of the public. 1990s were the years when the design of new products and development of national R&D activities emerged.

When the transformation in the Korean manufacturing industry in between 1970–2000 is reviewed, it is seen that the share in GDP of knowledge-based sectors had increased in time. In this process, the direction of transformation, which was from labor-intensive industries (such as food & beverage, textile & apparel) to capital-intensive industries before, changed into a move towards knowledge-based industries. Electrical and electronics products' percentage share in the GDP was 3.7 in 1970, and then this share increased to 10.4 in 1980, 14.6 in 1990, and 25.2 in 2000. This rise shows Korea's success of late development by learning-based technology transfer policies (Table 7.1).

Brands that are globalizing contributed to technological development and economic performance in South Korea's late-development process. Global brands, which are *Chaebol* companies such as Samsung, Hyundai, and LG, had played a decisive role in South Korea's knowledge-based technological development and industrialization process.

POSCO (Pohang Iron and Steel Company, Ltd) was operating in the Korean steel industry as a dominant firm. The government established the first integrated steel mill in Korea – POSCO in the late 1960s, early 1970s – which became one of the best-performing steel companies in the world a few decades later.[38] The steel industry developed rapidly following the establishment of POSCO in the Korea. The Korean steel industry became the catalyst and linchpin for a number of industries, such as automobiles, shipbuilding, containers, railroads, construction, and appliances, which complemented one another in a virtuous vicious circle of economic growth over the last three decades.[39] POSCO demonstrates the role and

[38] Mah (2007), p. 78.

[39] Shin and Ciccantell (2009), p. 176.

Table 7.1 Top 10 leading industries in Korea's manufacturing sectors (Percent of GDP)

	1970		1980		1990		2000	
Rank	Industries	Share	Industries	Share	Industries	Share	Industries	Share
1	Food & beverage	28.6	Textile & apparel	19.2	E&E products	14.6	E&E products	25.2
2	Textile & apparel	20.4	Food & beverage	19.0	Automobile	13.2	Chemicals	13.9
3	Chemicals	11.5	Chemicals	13.1	Food & beverage	12.9	Automobile	11.3
4	Automobile	9.1	E&E products	10.4	Chemicals	12.9	Basic metal	8.0
5	Paper & printing	5.5	Basic metal	6.7	Textile & apparel	11.5	Food & beverage	6.9
6	Nonmetallic mineral products	5.3	Automobile	6.1	Basic metal	9.0	Machinery	6.9
7	Coal & petroleum refinery	4.2	Coal & petroleum refinery	5.5	Nonmetallic mineral products	5.6	Textile & apparel	6.9
8	E&E products mineral products	3.7	Nonmetallic	5.3	Machinery	5.5	Fabricated metal products	4.8
9	Machinery	2.3	Paper & printing	3.9	Paper & printing	4.6	Paper & printing	4.3
10	Basic metal products	1.5	Machinery	3.7	Fabricated	3.8	Coal & petroleum refinery	4.2
	All manufacturing (% of GDP)	21.2	All manufacturing (% of GDP)	28.2	All manufacturing (% of GDP)	28.8	All manufacturing (% of GDP)	29.4

Source: Suh (2009), pp. 37, Table 1–4.
Note: Shares are of manufacturing value-added total. E&E = electrical and electronics.

importance of the state in economic development as one of the good and significant examples. Akkemik summarizes the government's role as follows[40]:

> ...during the Heavy and Chemical Industrialization Drive in Korea (1973–1979), the promoted industries were those the government saw as necessary to increase the self-sufficiency of the country in basic inputs (e.g., iron, steel, petrochemicals, and non-ferrous metals) and the technology-intensive sectors (e.g., shipbuilding, electronics, machinery, and equipment)...

Since 1973, South Korea has been building up and expanding its shipbuilding industry, and for the last couple of years, Japan and South Korea share world leadership in shipbuilding. South Korea's "big three" shipbuilders, the Hyundai Heavy Industries, the Samsung Heavy Industries, and the Daewoo Shipbuilding & Marine Engineering, dominate global shipbuilding. The Hyundai Heavy Industries (HHI), a subsidiary, began building its first ship, a very large crude carrier, in March 1973. Less than a decade later, the HHI became the world's largest ship-builder, with cumulative deliveries exceeding 10 million dead-weight tons by 1984.[41] The HHI, which produced both ships and heavy machinery and equipment, have a significant role in Korea's late-industrialization. Hyundai's role also in the automobile industry had been differentiated as the Hyundai Motor Company (HMC), in time. The HMC developed by applying its own technology through learning-based technology transfer. Amsden describes HMC's technological learning process in the following way[42]

> ... Hyundai adopted a policy of *obtaining technology from several sources* rather than from single one (sometimes it turned to two different sources for the same technology). From 1974 to 1976, HMC acquired technologies for engine block design, transmissions, and rear axles *from Japan*; for factory construction, layout, an internal combustion engines *from England*; and for car design *from Italy*. Eighteen technology transfers took place before introduction of the first HMC model, the 'Pony.' The experience that allowed HMC to *absorb these technologies was gained through is technical assistance from Ford...*

Korean firms have started activities in Thin Film Transistor (TFT)–Liquid Crystal Display (LCD) and DRAM sectors in the second half of the 1990s, and then have been surprisingly rapid, becoming heavyweights in only a few years in the world markets in only a few years. Samsung Electronics, which was one of the major firms in Korea, was ranked first in the world TFT-LCD world market in 1998 just 3 years after the company began mass production of TFT–LCD goods. In 2005, Samsung Electronics and LG Philips LCD together commanded 42% of the world market

[40] Akkemik (2008), p. 85.

[41] Amsden (1989), p. 269.

[42] Amsden (1989), p. 175.

share for TFT–LCD. Park et al summarizes Korea's success in the TFT–LCD sector below using related literature[43]:

> ... *At the firm level*: Korea's success can be attributable to high production technology and technological independence, a unique propensity toward large scale investment and the first-mover strategy that enabled Korea to compete with the world leader, Japan, high vertical integration downstream and part of the upstream (e.g., color filter suppliers), and strategic alliances with the United States, Japan, China and ASEAN. ... *At the industry level*: The TFT–LCD industry has the so-called crystal cycle, which repeats the changeable nature of business on a regular basis and offers the opportunity of successful entry to new firms when the cycle is facing a downturn. Korea was simply the country entering in the second downturn period and becoming successful. ... *At the national level*: strong support of the Korean government has played an important role in Korea's success. Because the TFT–LCD industry requires an enormous amount of money, the Korean government provided chaebols, such as Samsung and LG, with direct financing, thereby easing their fund raising. ... *At the technology level*: Technology relatedness, such as the production process technology, between TFT–LCD and DRAM elevated the learning effects in TFT–LCD and allowed for the straightforward accumulation of advanced technology....

To sum up, Korean late-industrialization experience shows that learning-based technology acquisition has been decisive. Sectoral experiences represent the best examples such as Hyundai. Learning-based technology transfer policies that ensure transformation from imitation to innovation with outward-looking development strategy provide South Korea's differentiation from Asian tigers just as Asian tigers' differentiation from regional economies as a late late-comer.

Conclusion and Lessons for Late Late-comers

The basic lesson that could be learnt from South Korea's experience is related to development stability and communal motivation. After the Korean War, development stability, communal motivation, and sustainability were the basic factors in Korean's transformation from the war economy to a knowledge economy within nearly 40 years. These basic factors that shaped the economic factors directed and led the sustainable development path. In other words, this kind of social performance marked the beginning of the South Korea's industrial transformation, and this, in turn, enabled the country to compete with the industrialized economies within less than half a century.

The Korean experience shows that development stability and social motivation were the basis of late development. This basis was strengthened by a developmental state model. If we make an evaluation of the Confucian culture, a homogeneous society structure including the supportive effect of the poverty, it will be relatively easy to understand the structural transformation from a learner economy to a teacher economy. The developmental state has a great importance in the creation

[43] Park et al (2008), pp. 2856–2857.

of a system that provides knowledge production and the transformation of this knowledge by commercialization. This, in turn, contributes to the development of human resources, skills, and productivity in the Korean experience. On the one hand, as the developmental stability and communal motivation increased the sensitivity on education, this sensitivity brought on the development of science and technology with the help of the state-supported education and health policy. As a result, the supportive effect of these industrial and technological policies was the creation of a microperformance. As the support for the government-funded research institutes, such as the Korea Institute of Machinery and Metals (KIMM), the Electronics and Telecommunications Research Institute (ETRI), the Korea Research Institute of Chemical Technology (KRICT), the Korea Research Institute of Standards and Science (KRISS), the Korea Institute for Energy Research (KIER), the Korea Ocean R&D Institute (KORDI), was restructured, South Korea's name has begun to be mentioned together with the technology leaders like the USA and Japan.

As the learning-based technology acquisition was backed with investments in human capital accumulation, technological learning capability, increased absorptive capacity, and as they are sustained with development stability and supported with communal motivation, South Korea came to constitute an excellent example of how good results could be achieved. Learning-based technology transfer policies that ensure transformation from imitation to innovation with an outward-looking development strategy have provided South Korea with a differentiation from the Asian tigers similar to the differentiation of the Asian tigers from regional economies.

In the knowledge-based economic development process, the most important issue to be kept in mind is that technology transfer from an external source must be added to the national efforts for technological development, absorbed in a way that ensures technological learning, Otherwise, it will not be possible to use it either for technology transfer or for development. It will not provide any benefits either. As a result, the argument that technological development alone is not the "*holy savior*" for late late-comers should always be kept securely in mind.

Acknowledgments The author is grateful to Murat A. Yülek for his valuable comments, ideas, and support.

References

Akkemik KA (2008) Industrial development in East Asia: a comparative look at Japan, Korea, Taiwan, and Singapore. World Scientific, Singapore

Amsden AH (1989) Asia's next giant: South Korea and late industrialization. Oxford University Press, New York

Chang HA, Grabel I (2004) Reclaiming development: an alternative economic policy manual. Zed Book, London

Freeman C (1987) Technology policy and economic performance: lessons from Japan. Pinter, London

Freeman C (1988) Japan: a new institutional system of innovation? In: Dosi G, Freeman C, Nelson R, Silverberg G, Soete L (eds) Technical change and economic theory. Pinter, London and New York

Freeman C (1989) New technology and catching up. The European Journal of Development Research 1(1):86–99

Freeman C, Soete L (1997) The economics of industrial innovation, 3rd edn. England, Pinter

Gerschenkron A (1962) Economic backwardness in historical perspective. Harvard University Press, Cambridge

Hamitoğulları B (1974) Teknoloji Transferinin Bazı Teknik Sorunları (Some Technical Problems of Technology Transfer), Türkiye Ekonomi Kurumu (derl.), *Teknoloji Transferi Sorunu ve Türkiye (Technology Transfer Problem and Turkey)*. Türkiye Ekonomi Kurumu Yayınları, Ankara, pp. 5–36

Hoekman B, Keith M, Maskus E, Saggi K (2005) Transfer of technology to developing countries: unilateral and multilateral policy options. World Development 33(10):1587–1602

Işık O (1981) Teknoloji Üretimi, Teknoloji Transferi (Technology Production, Technology Transfer) 2.Türkiye İktisat Kongresi, VI, Sanayi Komisyonu Tebliğleri, 2–7 Kasım 1981, İzmir

Kim C (2005) An industrial development strategy for Indonesia: lessons from the South Korean Experience. Journal of the Asia Pacific Economy 10(3):312–338

Kiper M (2004) Teknoloji Transfer Mekanizmaları ve Bu Kapsamda Üniversite-Sanayi İşbirliği (Technology Transfer Mechanisms and University-Industry Cooperation in this Context), TMMOB (derl.) *Teknoloji*. TMMOB Yayınları, Ankara, pp. 59–122

List F (1841) The national system of political economy (1942) (trans: Lloyd SS). Longmans and Green, London

Lundvall B (1988) Innovation as an interactive process: from user-producer interaction to National Systems of Innovation. In: Dosi G, Freeman C, Nelson R, Silverberg G, Soete L (eds) Technical change and economic theory. Pinter, London and New York

Lundvall B (1992) Introduction. In: Lundvall BA (ed) National systems of innovation: towards a theory of innovation and interactive learning. Pinter, London

Mah JS (2007) Industrial policy and economic development: Korea's experience. Journal of Economic Issues 41(1):77–92

Nelson R (1988) Institutions supporting technical change in the United States. In: Dosi G, Freeman C, Nelson R, Silverberg G, Soete L (eds) Technical change and economic theory. Pinter, London and New York

Nelson R (1993) National innovation systems: a comparative analysis. Oxford University Press, New York

Nelson R, Rosenberg N (1993) Technical innovation and national systems. In: Nelson R (ed) National innovation systems: a comparative analysis. Oxford University Press, New York

OECD (1995) Canberra manual. OECD & ECSC-EC-EAEC, Paris

Öniş Z, Şenses F (2009) "Küresel Dinamikler, Ülkeiçi Koalisyonlarve Reaktif Devlet: Türkiye'nin Savaş Sonrası Kalkınmasında Önemli Politika Dönüşümleri" (Global Dynamics, Intra-country Coalitions and Reactive State: Important Policy Conversions in Turkey's Post-war Development) in *Neoliberal Küreselleşme ve Kalkınma: Seçme Yazılar* (Neo-liberal Globalization and Development: Selected Articles), Fikret Şenses (Ed.), Istanbul, İletişim Yayınları

Pack H, Saggi K (2001) Vertical technology transfer via international outsourcing. Journal of Development Economics 65:389–415

Pamukçu MT (2001) Teknoloji, Sanayileşme ve Türkiye: Quo vadimus? (Technology, Industrialization and Turkey: Quo vadimus?). Mülkiye Dergisi 25(230):77–118

Park T, Choung J, Min H (2008) The cross-industry spill-over of technological capability: Korea's DRAM and TFT–LCD industries. World Development 36(12):2855–2873

Pavitt K (2001) Technology transfer among the industrially advanced countries: an overview. In: Lall S (ed) The economics of technology transfer. Edward Elgar Publishing Limited, UK and US

Posner MV (1961) International trade and technical change. Oxford Economic Papers 13 (3):323–341
Radosevic S (1999) International technology transfer and catch-up in economic development. Edward Elgar, Chetenham
Schumpeter JA (1942) Capitalism. Harper Torchbooks, Socialism, and Democracy, New York
Shin K, Ciccantell PS (2009) The steel and shipbuilding industries of South Korea: rising East Asia and Globalization. Journal of World Systems Research 15(2):167–192
Suh JH (2009) Development strategy and evolution of Korea's innovation system. In: Suh JH (ed) Models for national technology and innovation capacity development in turkey. Korea Development Institute, Korea Ministry of Strategy and Finance, Turkey
Teece DJ (1977) Technology transfer by multinational firms: the resource cost of transferring technological know-how. The Economic Journal 87(346):242–261
Tiryakioğlu M (2006) *Araştırma Geliştirme-Ekonomik Büyüme İlişkisi: Seçilmiş OECD Ülkeleri Üzerine Uygulama* (The Relationship between R&D and Economic Development: An application on Selected OECD Countries). Masters thesis, Afyon Kocatepe Üniversitesi, Sosyal Bilimler Enstitüsü, Yayımlanmamış Yüksek Lisans Tezi
Tiryakioğlu M (2008) Gelişmekte Olan Ülkelerin Çikmazi: Beşeri Sermaye Yoksulluğu (Dilemma of Developing Countries: Human Capital Poverty). Ege Akademik Bakiş 8(1): 321–339
Vogel E (1991) The four little dragons: the spread of industrialization in East Asia. Harvard University Press, Cambridge
Weiss L, Hobson J (1999) *Devletler ve Ekonomik Kalkınma: Karşılaştırmalı Bir Tarihsel Analiz* (States and Economic Development: A Comparative Historical Analysis), (Çev. Kıvanç Dündar), Ankara, Dost Kitabevi Yayınları
World Bank (1999) World development report: knowledge for development. Oxford University Press, USA
World Bank (2008) Global economic prospects-2008, technology diffusion in the developing world. The World Bank Publications, USA
Yülek M (1998) Asya Kaplanları: Sanayi Politikaları ve Kalkınma (Asian Tigers: Industrialization Policies and Development). Istanbul, Alfa Yayınları

Chapter 8
Technology Transfer in China: Analysis and Policy Proposals

Fuquan Sun and Wanjun Deng

Introduction

Over the past 30 years since the reform and opening up, the Chinese technology transfer system has been constantly developing and improving, and the system, including technology transfer subjects, technology transfer services organizations, and technology markets, has taken the initial shape. However, the Chinese technology transfer system is also faced with several problems. For example, the subjects lack initiative, the services organizations are low in the standard of service, the technology market has not developed a cooperation mechanism and the regulation system is incomplete. Thus, China will take stock of her own experiences and learn from the successful experiences abroad, while continuing to improve the technology transfer system.

Historical Evolution of the Chinese Technology Transfer System

Technology transfer is defined as systematic knowledge transfer on manufacturing products, application production methods, and supplying services, and it does not include buying and selling and leasing out merchandise in *the United Nations Draft International Code of Conduct on the Transfer of Technology* (Hualiang 2010). Technology transfer is an effective approach for the flow of technological achievements and it is beneficial to promote creation, diffusion, and value realization of technology. Technology transfer is an important component of the national innovation system and it is also a crucial part of international communication and cooperation in economy, science and technology, and trade.

F. Sun (✉) • W. Deng
Institute of Comprehensive Development (CASTED), Beijing, People's Republic of China
e-mail: sunfq@casted.org.cn; dengwj@casted.org.cn

M.A. Yülek and T.K. Taylor (eds.), *Designing Public Procurement Policy in Developing Countries*, DOI 10.1007/978-1-4614-1442-1_8,

Technology transfer system as a whole consists of the relevant government departments, technology transfer subjects, technology transfer services organizations, and technology markets. The government guides, encourages, and serves technology transfer by improving the market system and environment and by formulating the policies (Zhijun 2008). The technology transfer subjects consist of technology suppliers and technology users (receivers), and they are individuals and organizations such as R&D departments and technological enterprises. The technology transferred comprises patents, technical know-how, R&D achievements, etc. (Huili 2006). The technology transfer services organizations can accelerate achieving technology transfer through supplying various services, such as services for technology brokerage, start-up, technology integration and management (Xuhong and Zheng 2008).

With reform of China's science and technology system and establishment of the innovation system, the Chinese technology transfer system has roughly gone through five stages of development.[1]

Elementary Stage (1978–1984)

As the reform and opening up started in China, the scientific and technological activities took a historic turn toward economic construction and the technology transfer system also began to face the market. In March 1978, the National Conference on Science approved *the Outline of the Program for the State Science and Technology Development (Draft)* and insisted to promote application and extension of technology achievements and focus on the intermediate links from scientific research to production. In October 1980, the State Council promulgated *the Provisional Rules of Developing and Upholding Socialist Competition* and affirmed that the important technology achievements such as invention needed to be transferred with compensation for the first time. In November 1984, the State Council produced *Technology Transfer Ordinance* and formally introduced the concept of the technology market.

Exploration Stage (1985–1991)

Led by the guideline that science and technology must face economic construction and that economic construction must depend on science and technology, the polices on technology transfer were actively explored and the technology transfer system began to take shape. *The Decision on Reform of Science and Technology System*

[1] The stages refer to *30-Year Reform and Opening up of China Science and Technology.*

promulgated by the Chinese Communist Party Central Committee in 1985 and *the Decision on some Important Issues Concerning Deepening Reform of Science and Technology System* promulgated by the State Council in 1988 became basic polices to promote application and production of technology achievements. The government opened research institutions and extended their decision-making powers; encouraged the institutions of research, education, and design to cooperate with production factories; encouraged the research institutions for technological development to enter into enterprises; strengthened the enterprises' abilities on technology absorption and development; and built high and new-tech industrial zones to support private enterprises. Many laws and regulations, such as *Interim Measures of Technology Transfer*, *the Patent Law*, and *the Law on Technology Contracts* were promulgated to promote the opening of the technology market, technology achievement commoditization, and the flow of qualified people. In 1991, the technology contractual turnover was 9.48 billion Yuan, 4.6 times more than in 1986.

Market-Oriented Stage (1992–1998)

As deepening of the reform on the market economy continued, the technology transfer system also depended in a major way on market regulation and needed to energetically explore the technology markets. In 1992, the State Council promulgated *the Ten-Year Plan for the Development of Science and Technology in the People's Republic of China and the Outline of the Eighth Five-Year Plan,* and pointed out that it would be necessary to reform the planning administration and build the market mechanism, including the technology market, as well as the market support factors such as funding, materials supply, human skills, and information. The government mobilized more technology staff to establish private technology enterprises, produced the plan on constructing technological centers within enterprises, supported high tech to form the equity as intangible assets and promoted innovation projects on the integration of production, teaching, and research. In April 1994, the State Science and Technology Commission and the State System Reform Commission issued *Some Opinions on Further Cultivating and Developing the Technology Market* and pointed out that more attention should be paid to develop technology exchange, establish many new types of organizations such as the engineering research center, the productivity promotion center, and the technology incubation center, while also creating intermediary institutions of technological trading and a fair, just, and open market order.

Deepening Stage (1999–2005)

As the research institutes were transformed toward enterprises and the market support factors developed, the modern technology market system and the technology transfer system were improved and, thus, the capacities on innovating technology and industrializing high-tech achievements were enhanced. The applied research institutes were transformed into enterprises and the public research institutes were reformed on the basis of being profitable or unprofitable, in order that the science and technology system and mechanism could be beneficial for the transfer and industrialization of the existing scientific and technological achievements into productive endeavors. The research institutes and the colleges and universities had to transform their service technology achievements toward the market. In December 1999, the Shanghai Technology Property Rights Exchange was established so that China would realize resource integration in the technology market, the financial market, and the property rights market. In addition, China began to encourage risk investment and set up venture funds for medium- and small-sized enterprises in order to promote technology driven industrialization.

Innovation-Oriented Stage (2006–Now)

The technology transfer system began to serve indigenous innovation and building an innovative country by strengthening the main firms and by integrating production, teaching, and research. At the end of 2005, the State Council promulgated *the Outline of the Program for the State Long-term Science and Technology Development (2006–2020)*. It confirmed the guidelines "independent innovation, spanning on the key, supporting development, guiding future," and raised the general objective of constructing an innovative country, thus, signaling a new stage of technology transfer system. The government put forward more than 60 measures that guaranteed and supported indigenous innovation and technology transfer in areas such as the fiscal system, taxation, banking, industries, government purchases, introduction and absorption, and intellectual property rights. The market support factors such as funding, materials, human skills, and information needed to be further improved, the technology market be actively developed, innovative elements be gathered toward the corporations, technology achievements transformation and technology transfer be promoted, and the system of identification and registration of state technology contracts on the internet be started in order to dynamically manage and analyze the state technology contracts. In 2008, the total value of technical contracts signed on the state technical markets reached 266.523 billion Yuan.

Progress of the Chinese Technology Transfer System

The technology transfer system is a product of the reform on the science and technology system, and it is a significant innovation of the system and mechanism in the course of developing the Chinese socialist market economy. After 30 years of cultivation and growth, China has basically developed a technology transfer system and scored a number of achievements.

Initially Establishing a System of Policies, Laws, and Regulations on Technology Transfer

China has developed a legal system and a management system of technical contracts whose major frame includes the state laws and regulations on scientific and technological development, technology transfer, and contracts. There are 32 provinces, autonomous regions, municipalities, and cities that have specifically made administrative rules on the technical market. To promote the flow of the technical elements and technology transfer, on the basis of the previous polices covering business tax remission on technology development and transfer, the new tax incentives aiming to encourage home-ground innovation and promote technology transfer were issued (Gang et al. 2008).

Forming a Pattern of Diversified Technology Transfer Subjects

Before the reform and the opening up process, most Chinese technical resources were in research institutes, but now innovative subjects have been diversified to research academies and institutes, colleges and universities, and enterprises. In 2008, the expenditure on R&D was 461.60 billion Yuan and there were 1.97 million R&D employees. According to the percentages of every innovative subject in Figs. 8.1 and 8.2, enterprises, governmental research institutes, and colleges and universities are playing important roles. Meanwhile, the enterprises' position on innovative subjects is enhanced.

Technology users also comprise many types of enterprises, such as self-employed enterprises, enterprises in foreign countries, enterprises with foreign investment, and enterprises with investments from the mainland, Hong Kong, Macao, and Taiwan. Figure 8.3 indicates the percentages of transaction values of technology users in the technical market. Except enterprises with foreign investment, other types occupy important positions. Especially, the increase in the proportion of self-employed enterprises can to some extent reflect the effect of the state support for small businesses.

Rapidly Building Technology Transfer Services Organizations

In China, technology transfer services organizations have fast grown from nothing. In the 1980s, the first batch of transfer services organizations was built. After a 30-year or so development, 189,800 services organizations have been established. By the end of 2008, there had been more than 200 standing technology markets and nearly 40 transaction agencies with technological property rights and financial capital (Gang et al. 2008). There had been 1,532 state productivity promotion

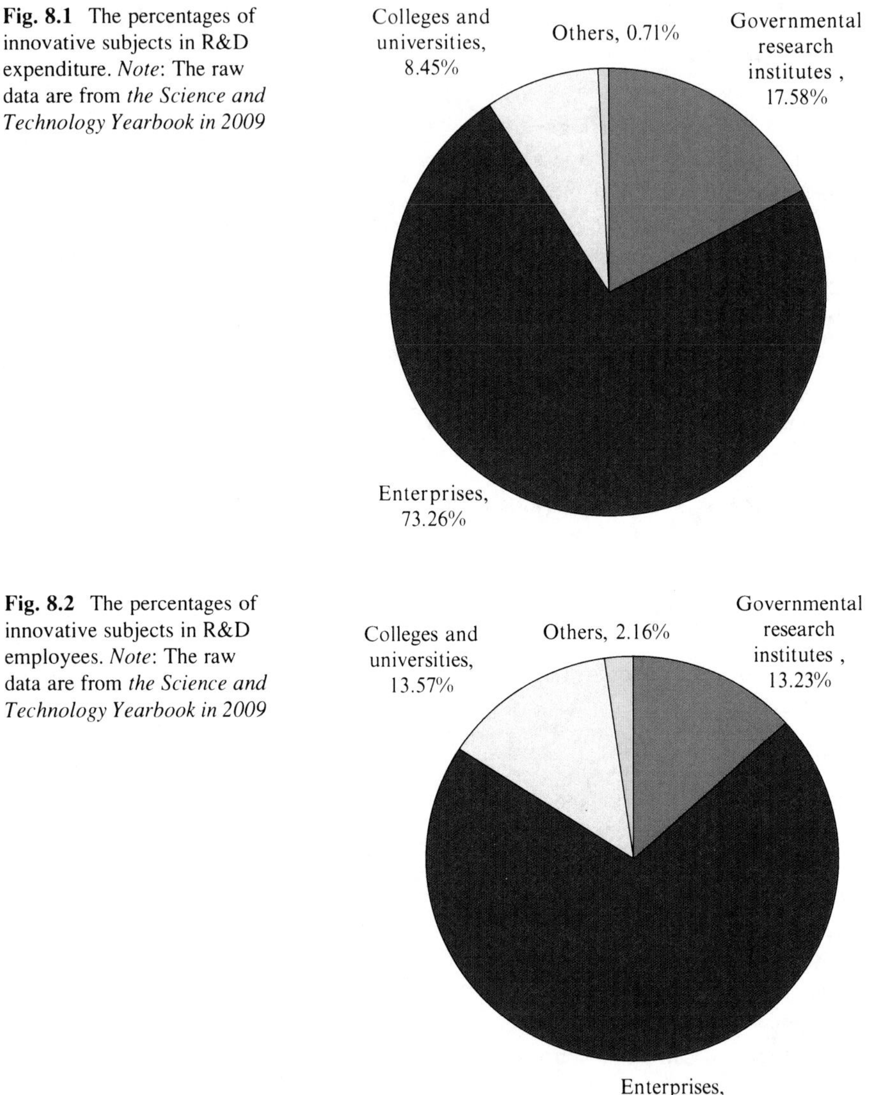

Fig. 8.1 The percentages of innovative subjects in R&D expenditure. *Note*: The raw data are from *the Science and Technology Yearbook in 2009*

Fig. 8.2 The percentages of innovative subjects in R&D employees. *Note*: The raw data are from *the Science and Technology Yearbook in 2009*

centers which supplied services for small- and medium-sized enterprises, with an income of 3.04 billion Yuan, and 164 centers were at the national demonstrated level. There had been 674 technological business incubators, and 228 business incubators were at the national level. 32,370 enterprises had been incubated and 44,832 were in the process of being incubated. In 69 state science and technology parks in colleges and universities, there had been 1,055 public service organizations for business start-up and 247 industrialization service platforms. 2,979 enterprises had been graduated. The firms in the parks had undertaken 2,918 projects of various

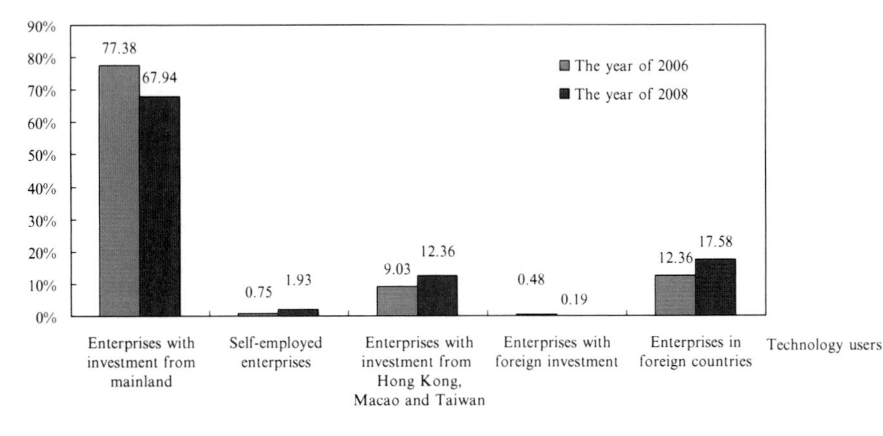

Fig. 8.3 The percentages of transaction values of technology users in the technical market in 2006 and 2008. *Note*: The raw data are from *the Science and Technology Yearbook in 2009*

levels and sorts and had applied 4,454 patents.[2] The state promoted technology transfer by approving 134 demonstrated transfer institutions at the national level to take the lead, developing the regional alliances for technology transfer to promote technology cooperation in a certain region, and building the innovation relays to support multi-national technology transfer, international cooperation of production, teaching, research, and technology innovation for small- and medium-sized enterprises.

Significant Increase in the Scale of Technical Contracts

The contracted volume in the Chinese technical market was 700 million Yuan at the early period of the opening of the market, but by 2008 it reached 266.52 billion Yuan. It took 17 years to reach 100 billion Yuan of the technical contracted value, but from 100 to 200 billion Yuan, it took only 4 years. Figure 8.4 shows the number of registered technology contracts and their transaction values during 2000–2008. Although the number of registered technology contracts had a fluctuation, the transaction value indicated a favorable development, and its increasing speed had a growing trend. In addition, the average transaction value for each technology contract was 2.7 billion Yuan in 2000 and then reached 11.78 billion Yuan in 2008. The average annual growth rate was 20.21%, and it reflected the increase of the transaction volume of each contract. All the above reflect that Chinese technical market is in a period of fast development.

[2] The raw data come from *the Report on Science and Technology Development in China in 2008*.

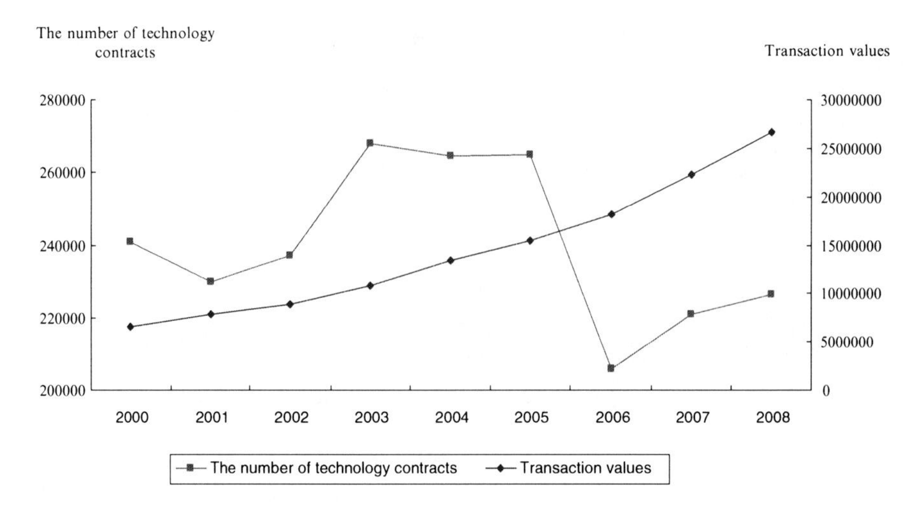

Fig. 8.4 The number of registered technology contracts and their transaction values during 2000–2008. *Note*: The raw data are from *the Science and Technology Yearbook in 2009*

Problems in the Chinese Technology Transfer System

In China, technology transfer system has been a weak component of the state innovative system and the commercialization and industrialization of technical achievements have been far from enough. According to the reports, the technical achievements which were in long-time use in production are smaller than 20% and the achievements which passed on to industrialization occupy about 5% of the total. In comparison, the rate of technical achievement transfer from foreign countries is more than 50%. If the transferred achievements are reflected by the technology contracts, and technical achievements are reflected by the invention patents, the ratio of the above measures shows the degree of commercialization of technical achievements, which is shown in Fig. 8.5. Since the entry into the twenty-first century, the figures have obviously descended. It further reflects the unhealthy development in the commercialization of technical achievements.

The major concrete problems in the Chinese technology transfer system are listed as follows:

Lack of Initiative in the Subject of Technology Transfer

Technology suppliers seem less energized in studying the technology that can be transferred and technology users also lack enthusiasm to employ independent R&D techniques. The major reasons are lack of incentives and limited funds for technology transfer.

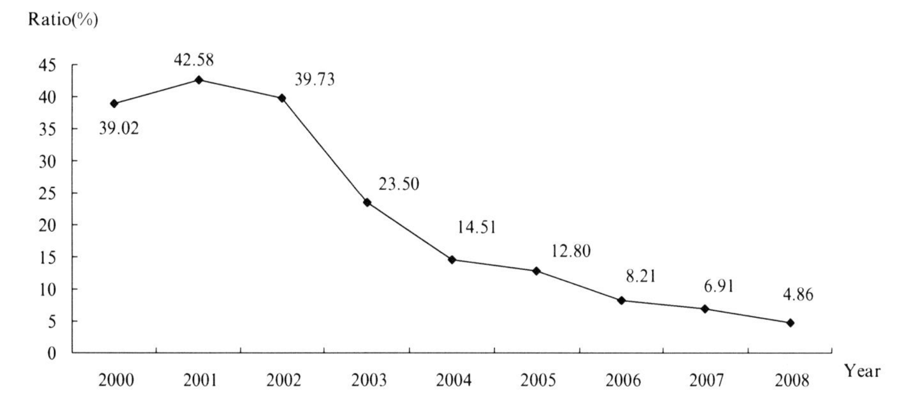

Fig. 8.5 The ratio between the transferred achievements/the technology contracts and technical achievements/invention patents. *Note*: The raw data are from *the Science and Technology Yearbook in 2009*

The current incentive and evaluating methods cause the technology transfer subjects to lack enthusiasm. For technology suppliers, the evaluation system focuses more on advanced technology and pays little attention to the industrialization of the technology. Thus, the technical institutes suffer from the lack of motive power for technical industrialization (Guanghui and Wei 2009). In contrast, in western countries, when the project is established, the feasibility of the technology transfer has been demonstrated and the technology transfer effect has been made an important component of examination. The firms that are technology receivers have no special polices and tax preferences relating to indigenous innovative end-results. Meanwhile, since most of the technical achievements by the research institutes are realized at the laboratory stage, the companies need to undertake a lot of investment and establish technical ventures for the realization of the second or third development phases (Guanghui and Wei 2009). Thus, the companies prefer the matured technologies to the transferred ones.

The technology transfer necessitates high venture, high investment, and is spread over a long cycle, so that the shortage of the funds is decisive for technology transfer. This factor also results in the fact that the technology transfer subjects cannot actively participate in the transfer. Figure 8.6 shows that in China most technical funds come from the government and companies, with a share of over 80%. In the total, meanwhile, the research institutes have almost no funds for transformation and transfer, and the funds in monetary institutions and social funds in other forms make up just a small proportion. When the distribution of the corporate funds is further analyzed, it is seen that in the Chinese enterprises the R&D expenditure is 0.5% of the business income, a share that is greatly different from that in foreign countries, where it reaches 6%. Therefore, when the monetary and social funds do not effectively come into the technology transfer process, the technology transfer subjects end up with having no capacities to face the great ventures or raise funds to realize the transfer of technology in question.

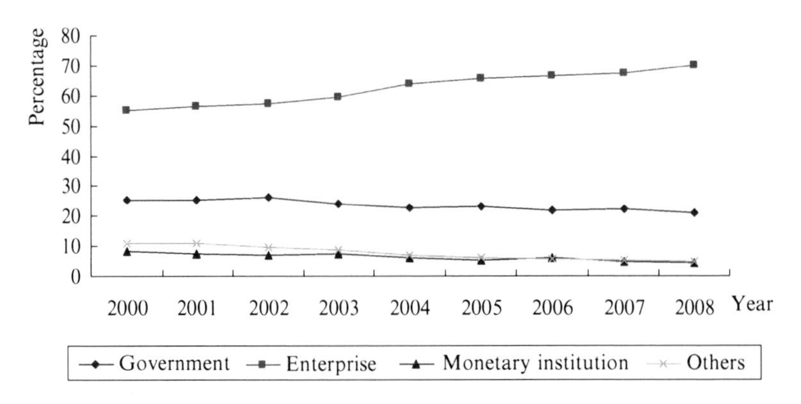

Fig. 8.6 The percentage of each subject in the funds raised for the scientific and technological activities in China. *Note*: The raw data are from *the Science and Technology Yearbook in 2009*

Technology Transfer Services Organizations' Weak Service Capacities

In the recent years, the technology transfer services organizations have experienced a certain amount of development from nothing, although, the organizations' and the employees' service capacities are still weak.

On the one hand, there are various kinds of services organizations, but their service capacities cannot satisfy the market. Most services organizations, with their small scales, are in the preliminary stage, so they can hardly supply timely and correct information in their limited service range and they face difficulties in meeting the practical demands of large-scale technology transfer. Also, the service system for the small- and medium-sized enterprises needs to be improved. Although China has established many institutes and departments, such as the productivity promotion center which specially serves the small- and medium-sized enterprises, these still need more services organizations to help them obtain information and follow the developing trends.

On the other hand, China is short of the high-level professional employees needed in technology transfer. An outstanding technical broker is the versatile employee with the qualities and capacities of technical specialists, entrepreneurs, and sociologists. In the developed countries, such as the USA and Japan, most professional employees in technology transfer have master's degrees and doctorates, but in China most are experienced qualified persons, but in limited numbers. As a result, according to the Chinese technical market association, in 50,000 services organizations with 1.3 million staff, only a few persons were able to pass authentication (Guanghui and Wei 2009).

Technical Market's Deficiency in the Cooperation Mechanism

All the technology transfer institutes have a loose structure and lack a cooperation mechanism, so the ability of the networking combination is hard to form, which leads to the lack of an all-round, timely, and correct service. In China, most technology transfer institutes have built their information systems, such as the databases of technical achievements, technical demands, technical experts, and technical enterprises, just based on the needs and realities of their own districts, resources, and businesses. For example, colleges and universities establish the state and local technology transfer centers and in order mostly to transfer their own technology achievements, without any division of labor among them and a cooperation atmosphere. Thus, the individual institutes end up having only limited abilities for technology transfer and are able to complete only a small part of the work needed. Also the work is hard to divide up into parts and since the cooperation mechanism for technical integration, agent integration, information integration, and international communication integration has not also been formed, the same tasks are repeated again and again in the individual technology transfer institutes (Xuhong and Zheng 2008).

Incomplete Legal System of Technology Transfer

During the 30-year growth, China has passed many laws, rules, and regulations related to technology transfer and had a preliminary frame for a legal system of technology transfer. However, in the area of implementing the policies, the legal system is still incomplete. First, the current laws and regulations are dispersive, without a set of systematic laws in relation to technology transfer at the state level. Second, although the principal laws and regulations are in place, there are fewer detailed measures, so that the subjects are unclear in division of the work and the benefits, and the ownership of the intellectual property rights.

Policy Proposals Toward Improving the Technology Transfer System in China

In order to promote technology transfer and improve the technology transfer system, China needs to learn from the successful experience at home and abroad to be able to adopt measures to help solve the above problems. Some policy proposals are listed as follows:

Cultivating Technology Transfer Subjects' Enthusiasm

Governments at all levels can make an attempt to reform the system of evaluating and encouraging technology transfer subjects, in order to cultivate their enthusiasm and attract more financial institutions and other social institutions that would participate in the technology transfer process.

More technology transfer elements should be added to the evaluation of technology innovation subjects, but there should also be different evaluation methods according to specific research fields. Original research projects are sophisticated and they focus on the forward position. So they are hard to be judged in the short-run. Therefore, according to the experiences of Beijing Institute of Biological Sciences, one can invite experts in the same field to judge the original research projects over the long term.[3] Applied research projects can be evaluated through the technology transfer effect, quantities and qualities of papers and patents over a period of 2–3 years. Particularly, there should be a high weight placed on technology transfer, and then the researchers will pay more attention to technology transfer, making use of the research achievements already at hand. The reform on the system of encouraging the researchers can be mainly on reforming the material rewards in this area. One can refer to the experiences of the German Fraunhofer-Gesellschaft Association for the Promotion of Applied Research in this regard. They hold that a large number of personnel expenses are an important condition for keeping high R&D abilities at high levels. Thus, more than half of the cooperation funds are allocated to the staff and the researchers, and they can claim 20% of the profit which comes from the later application of patents.[4]

In the case of the enterprises, the government should encourage them to actively participate in technology transfer, first, through ways like financial allocation, establishing special funds, and providing loan subsidies, government supports to technology transfer services organizations so that they can establish pilot plants for scientific and technological achievements independently or cooperatively, in order to hatch scientific and technological achievements (Guanghui and Wei 2009). Second, through ways like governmental tax incentives, free consultation provided by governmental intermediary services organizations, and the reduction of housing rental fees provided by management committee of science and technology parks, the government should encourage companies to use the technology achievements independently researched in China. Third, enterprises should be encouraged to actively participate in studying issues and projects, and should take achievements transfer as an important project-checking-and-accepting index.

The government should encourage financial institutions and venture funds companies to invest in technology transfer in order to reduce the technical companies' risks in absorbing technology and to raise the initiatives of the technology transfer subjects. In view of the American experiences, the government can stimulate the development of venture capital and financial loans by confirming the system of tax incentives, financial subsidies, and credit guarantee for venture funds companies.

[3] The research institute of life sciences in Beijing innovative mechanism assembles overseas talents.

[4] An analysis on technology transfer system in France and Germany. A series of research reports on *Strategic Study on Establishing China Technology Transfer System Towards the 21st Century*.

In the late 1970s and early 1980s, in order to increase the personal venture investment, the American government reduced the capital gains tax rate from 49.5 to 20%. The result was that, in 1982, the net value of venture investment was seven times larger than that 12 years before. In addition, through policies and projects, the government can attract more social funds and foreign capital to join in the Chinese technology transfer drive.

Fostering and Cultivating Technology Transfer Services Organizations of High Standards

Since technology transfer services organizations should play a greater role in promoting technology transfer, the government needs to support and cultivate a number of high-quality services organizations and professionals.

The services supplied by the intermediary services organizations have to cover all aspects of technology transfer. Meanwhile, in order to meet the demands of the intermediary services, the technical market needs that the public institutions funded by the government and the profitable organizations funded by society must coexist and be complementary. In this connection, first, the services organizations should enrich and refine their services. These organizations not only simply connect technology transfer subjects to one another, but also offer them specific suggestions on important strategic decisions such as development directives, technology transfer forms, etc. So the Chinese services organizations must continue to refine their present services and provide more specific consulting services as well.

The services organizations can take the experiences of Fraunhofer-Gesellschaft as a reference in order to provide information about the development trends so that the technical enterprises can accurately gauge the market orientation, and realize better developments in the more important technology fields. Second, the qualities of the services organizations should include both public and profitable realms. For the small- and medium-sized enterprises, which have no profitability, there should be ways to provide them with cheaper technology consultation. Therefore, the government should establish a number of public technology transfer services organizations, like OSEO in France and service agencies funded by the German government,[5] mostly directed to the needs of the small- and medium-sized enterprises. In addition, for some large-scale technology transfers that need a larger range of information and more strategic guidance, the government needs to cultivate a number of powerful technology transfer services organizations to provide comprehensive and high-level services.

---

[5] An analysis on technology transfer system in France and Germany. A series of research reports on *Strategic Study on Establishing China Technology Transfer System towards the 21st Century*.

High-level technology transfer professionals should be cultivated by both universities and the market, and their staff should be encouraged to learn with a spirit of initiative guiding them. First, an education system with high academic qualifications in the field of technology transfer should be established. According to the specialized demands of technology transfer, the postgraduate education with scientific courses and training schemes on technology transfer can be set, so as to cultivate high-quality, internationalized, and versatile technology transfer personnel with the spirit and capacity for technological innovation. Reportedly, Beijing Industrial University started to recruit in 2008 software engineering masters and graduates with an orientation in the direction of technology transfer, a good initiative for the high education system of China's technology transfer process. Second, technology transfer seminars, a kind of nondegree education, can be organized to train technology transfer talents. Technology market management offices in all provinces and cities can learn from that in Shanghai in this regard. They can increase the training courses of technology brokers and offer certificates of practice. Third, a vocational qualification system related to technology transfer can be established and developed. The government can encourage the management of technology broker resources and promote vocational qualification in Bohai Rim region, Yangtze River Delta, and other areas, greatly improving the service levels and professional ethics of the technology brokers (Guanghui and Wei 2009).

Constructing the Cooperation Mechanism in the Technical Market

By constructing the technology transfer alliances among countries, areas, and industries, and establishing various ways of technology transfer, it would be possible to integrate the present resources and realize the cooperation needed in the technical market.

The technology transfer alliance is an effective way to integrate various intermediary services organizations in the technical market, especially the medium and small ones. For one project, it is possible to assemble the corresponding technology transfer organizations to complete the projects at higher levels of quality. The state technology transfer alliance works to coordinate and instruct all sorts of the state technology transfer organizations in order to realize the cooperation and integration of all the resources. The regional technology transfer alliance is the center of regional technology transfer. It will undertake the responsibility of transferring the technical achievements invested by the country, as well as by the region. The industry technology transfer alliance can comprise the leaders, organizers, and coordinators of technology transfer in some sectors.

In order to better promote the cooperation between the technology transfer subjects, besides traditional technology exchanges, new ways which would be beneficial for technology creation and transfer, can be developed. These would include cooperation in production-study-research, such as in the cases of the Joint Research Institute established by Shanghai Baogang Steel Group and Shanghai JiaoTong University; building a new firm or a branch institute, such as the National

Heavy Machinery Research Institute – Yanshan University Branch built by the National Heavy Machinery Research Institute Co. Ltd. and Xi'an Jiaotong University, Northwestern Polytechnical University and many other universities, and Shanghai Modern Chinese Medicine Technology Co. Ltd., built by Shanghai University of TCM and Shanghai Xinhangning Group, which undertakes the development and industrialization of new medicinery[6]; enterprise technical transfer bases, such as the two pilot plants and some specific technology industrialization bases built by National Heavy Machinery Research Institute Co. Ltd.[7]; enterprises' joint development, such as the research body of the integrated circuit with a very large scale in Japan (Xinxin 2008); the industrial cluster, such as the Suzhou IT industry cluster and Haidian Science and Technology Park.

Establishing and Improving the System of Policies and Legislation

In the process of establishing and improving the technology transfer system, the government should play an important role in providing guidance and service by establishing and improving the system of policies and legislation.

First, on the basis of the existing laws and regulations related to technology transfer and the objective requirements of the development of the Chinese technology transfer, the government should draw up and promulgate the state legal documents targeted toward technology transfer as soon as possible. Government, universities, research institutes, companies, and services organizations can confirm their rights, obligations, and interests in technology transfer and can divide the work rationally in order to increase efficiency. Second, after publishing guiding principles and policies, both the national and local governments should formulate specific measures based on specific conditions in order to insure having the foundations of implementation at hand. Third, besides continuing to encourage scientific research departments to develop indigenous innovation, the government should strengthen its support toward the technology receivers and technology transfer services organizations in the areas of finance and taxes. Also, local governments can establish special funds for technology transfer services, for supporting technology transfer services organizations in order to enhance the capacity of the professional services. Fourth, the intellectual property system must be promoted within a short period. Technology transfer has inseparable relations with management and operation of intellectual property and, thus, a perfect system can guarantee intellectual property transfer among different subjects.

[6] Inspiration from the technology transfer system in Yangtze River delta zone. A series of research reports on *The Study on the Mechanism and Policy of Technology Transfer from University to Industry*, 2007.

[7] Very large scale research unions on the integrated circuit in Japan: A pioneering undertaking about the integration of government, production, teaching, and research, *A Selection of Independent Innovation Cases*, pp. 211–216.

References

Gang W, Yong S, Laiwu Z (2008) 30-Year reform and opening up of China science and technology. Science Press, Beijing

Guanghui W, Wei W (2009) Status, problems and suggestions for China technology transfer. Taiyuan Sci Technol 11:4–7

Hualiang F (2010) Science and technology achievement transformation and technology transfer: a discrimination of the two terms. Sci Technol Manage Res 10:229–230

Huili L (2006) Status and development suggestions for China technical transfer. Enterprise Reform Manage 1:30–31

Xinxin W (2008) Development trend and corresponding measures of China technology transfer. Sci Technol Manage Res 12:88–90

Xuhong D, Zheng Z (2008) Status and measures for China technical transfer services organizations. Sci Technol Ind China 5:78–81

Zhijun L (2008) The role of government in promoting technology transfer. Chin Univ Technol Transfer 4:92–93

Chapter 9
Dependence on Imported Inputs and Implications for Technology Transfer in Turkey

K. Ali Akkemik

Introduction

Trade and macroeconomic policies in Turkey evolved from import substitution to export promotion and liberalization of commodity and capital markets after 1980. During the 1980s and 1990s, Turkey's exports and imports and their shares in GDP demonstrated an increasing trend. The share of exports in GDP increased from 4.2% in 1980 to 20.3% in 2005 and that of imports rose from 11.4 to 32.2%. Import liberalization was accomplished during the second half of the 1990s and at around the same time direct price support for exports was abolished.

Furthermore, export promotion was driven by trade liberalization and wage suppression during the period 1980–1989. Subsequent capital account liberalization (1989–1994) completed the integration of the economy with global markets, which attracted short-term foreign capital. The resulting crisis in 1994 revealed a fragile economic structure amidst the transformation of the economy with liberalization of trade and foreign capital flows. This was worsened by external shocks (Asian crisis and earthquake) during the late 1990s. The period 1995–2000 was characterized by high inflation, larger trade deficits, and increasing indebtedness of the government Therefore, the government commenced a stabilization and disinflation program with the IMF support in 2000, which resulted in two financial crises in 2000–2001. After the crisis, institutional and macroeconomic reforms coupled with relative political stability helped Turkey achieve macro-stability and attract large inflows of foreign investment.

In a popular debate about the effectiveness of trade liberalization and export promotion after 1980, most studies argued that export-oriented policies in Turkey were not able to sustain growth, and, moreover, that that the liberalization strategy lacked a vision of industrial restructuring. Cross-country studies generally conclude

K.A. Akkemik (✉)
Kadir Has University, Istanbul, Turkey
e-mail: ali.akkemik@khas.edu.tr

M.A. Yülek and T.K. Taylor (eds.), *Designing Public Procurement Policy in Developing Countries*, DOI 10.1007/978-1-4614-1442-1_9,

that increasing openness brings about productivity gains through technological spillovers and externalities (e.g., Sachs and Warner 1995; Edwards 1998). Contrary to the expectations of this line of thinking, Boratav et al. (1996) criticized the structural adjustment policies after 1980, and claimed that the main reasons behind macroeconomic instability during the post-1980 era were weaknesses in the fiscal system and premature external liberalization. Yeldan (1989) argued that an industrial development strategy that emphasizes domestic demand expansion accompanied with selective export promotion was the most viable policy in the long run. Boratav et al. (2000) found that trade liberalization was insufficient to enhance competition in manufacturing industries, as evident from rising mark-up rates and the increasing concentration of the manufacturing sector. They also argued that the cost of export-led growth (i.e., subsidies and other forms of assistance allocated to export industries) was large in the sense that these industries could not generate productivity gains.

The aim of this chapter is twofold: (1) to examine the import dependence of Turkish exports using input–output tables and (2) to offer policy recommendations for technology structure of exports and technology transfer using the findings of the analyses. The chapter is organized as follows: section "Import Dependence of Exports in Turkey from the 1970s to 2000s" analyzes import dependence of exports. Section "Implications for Technological Content of Exports, Technology Policies, and Technology Transfer" raises issues about technology structure of exports and technology transfer. Finally, section "Conclusion: Technology Transfer and Policies to Encourage Innovation" summarizes the importance of import dependence of export and its implications for technology transfer.

Import Dependence of Exports in Turkey from the 1970s to 2000s

Recent studies have examined the dependence of exports and production in Turkey on imports using officially published input–output (I-O) tables (Günlük-Senesen and Senesen 2001; Senesen and Gunluk-Senesen 2003a, b; Esiyok 2008). The studies used similar I-O methods and argued that imported input requirement in the economy, in general, and export-oriented sectors, in particular, increased over time. In other words, the export spurt after 1980 was realized at least partially through imports of material inputs. Such findings cast doubt on the effectiveness and sustainability of the export-oriented strategy and trade liberalization in Turkey during the 1980s and 1990s.

In this section, the import dependence of Turkish exports using I-O tables are examined for 1979, 1985, 1990, 1996, and 2002, employing techniques similar to those used by a number of other researchers (Bulmer-Thomas 1982; Kim 1990; Gunluk-Senesen and Senesen 2001; He and Zhang 2010).

Method of Analysis

The standard Leontief input–output model postulates that the amount of a sector i's output necessary for the production of another sector j's output (X_{ij}) is proportional to sector j's output (X_j). This proportion is translated into the well-known input–output coefficients (a_{ij}), which is described as follows:

$$X_{ij} = a_{ij}X_j \tag{9.1}$$

The Leontief Model measures the impact of a change in the final demand (F) of a sector on all sectors as in the following equation:

$$X_i = \sum_j X_{ij} + Y_i = \sum_j a_{ij}X_j + F_i \tag{9.2}$$

Suppressing industry subscripts, (9.2) can be written in matrix notation as:

$$X = AX + F \tag{9.3}$$

A, X, and F are input–output coefficient matrix, intermediate demand vector, and final demand vector, respectively. The solution to (9.3) yields the multipliers that measure the impact of changes in the final demand on sectoral output levels:

$$X = (I - A)^{-1}F \tag{9.4}$$

The inverse matrix $(I - A)^{-1}$ is the multiplier matrix. It measures direct and indirect impacts of final demand on sectoral output. F vector consists of final demand for domestic activities (public and private consumption and investment), export demand, and imports of goods and services. Replacing the competitive input–output coefficient matrix (A) with the domestic input–output coefficient matrix (A^{D}), the impact of the final demand for domestic activities on the output levels of domestic sectors (X^{D}) is found as follows:

$$X^{\mathrm{D}} = (I - A^{\mathrm{D}})^{-1}F^{\mathrm{D}} \tag{9.5}$$

F^{D} is the final demand for domestic products. The domestic input–output coefficient matrix (A^{D}) is calculated by subtracting the imported input matrix (A^{M}) from the A matrix.

The changes in final demand in the economy (i.e., the F vector), especially exports, bring about changes in the industrial structure. Due to differences across industries in terms of dependence on imported inputs, changing industrial structure, in turn, stimulates changes in the imports of intermediate inputs. Some export sectors may be relatively more dependent on the imports of intermediate inputs. In this case, changes in exports in those industries will be correlated with the changes in the imports of intermediate inputs and their prices. Therefore, it is important to analyze import dependence. Declining import dependence implies increasing import substitution.

The I-O tables provide the matrix of imported input coefficients (A^{M}). Imports induced by a change in final demand for domestic products (M^{D}) are found as follows:

$$M^{\mathrm{D}} = A^{\mathrm{M}}X^{\mathrm{D}} = A^{\mathrm{M}}(I - A^{\mathrm{D}})^{-1}F \tag{9.6}$$

An increase in export demand also induces demand for imported inputs (M^{E}):

$$M^{\mathrm{E}} = A^{\mathrm{M}}X^{\mathrm{E}} = A^{\mathrm{M}}(I - A^{\mathrm{D}})^{-1}E \tag{9.7}$$

Using M^{E}, the impact of induced imports for a unitary increase in exports can be quantified. The resulting figures are import multipliers for exports.

We define three versions of import dependence (ID), following Kim (1990, p. 64). Import dependence of domestic demand (ID^{D}) is computed as the ratio of import (M^{D}) to output induced by domestic demand (X^{D}):

$$\mathrm{ID}^{\mathrm{D}} = M^{\mathrm{D}}/X^{\mathrm{D}} \tag{9.8}$$

Import dependence arising from export demand (ID^{E}) is calculated in a similar fashion as the ratio of imports to output, both generated by exports:

$$\mathrm{ID}^{\mathrm{E}} = M^{\mathrm{E}}/X^{\mathrm{E}} \tag{9.9}$$

Imports induced by intermediate input demand (M^{N}) are calculated using the vector of the coefficients of imported intermediate inputs which are denoted as N:

$$M^{N} = NX^{\mathrm{D}} = N(I - A^{\mathrm{D}})^{-1}F \tag{9.10}$$

Import dependence of intermediate inputs (ID^{N}) is then the ratio of imports generated by intermediate inputs to output generated by exports:

$$\mathrm{ID}^{N} = M^{N}/X^{\mathrm{E}} \tag{9.11}$$

Data

The data are obtained from the Turkish Statistical Institute's Input–Output tables published for the years 1973, 1979, 1985, 1990, 1996, and 2000. The number of sectors is 64, for the first 4 input–output tables. However, the number of sectors increases to 98 for 1996, and then decreases to 49 for 2002. These sectors are aggregated to 31 broad sectors as shown in Table 9.1. Eighteen of these sectors (05–18 and 20–23) are manufacturing industries. These sectors are also classified into six different product groups based on their resource and R&D intensity.

Table 9.1 Sectoral aggregation

		Sector codes		
Number	Sector definition	1979, 1985, 1990	1996	2002
Raw material-intensive products				
01	Agriculture	01–02	01–05	01
02	Forestry	03	06	02
03	Fisheries	04	07	03
04	Mining and quarrying	05–10	08–12	04–08
05	Coal and oil refining	32–33	38	17
Labor-intensive products				
06	Textiles	20–21	26–27	11
07	Wearing	22–24	28–32	12–13
08	Wood	25	33–34	14
19	Furniture	26	67	30
10	Paper	27	35	15
11	Printing	28	36–37	16
12	Other manufacturing	49	68	31
Capital-intensive products				
13	Food and beverages	11–18	13–24	09
14	Tobacco	19	25	10
15	Rubber and plastics	34–35	44–45	19
16	Nonmetallic minerals	36–38	46–49	20
17	Basic metals	39–40	50–51	21
18	Fabricated metal products	41	52–54	22
19	Electricity, gas, and water	50–51	69–71	32–33
Easily imitable R&D-intensive products				
20	Chemicals	29–31	39–43	18
21	Nonelectrical machinery	42–43	55–57	23
Difficultly imitable R&D-intensive products				
22	Electrical machinery	44	58–61	24–27
23	Transport equipment	45–48	62–66	28–29
Services				
24	Construction	52–53	72	34
25	Wholesale and retail trade	54	73–75	35–37
26	Hotels and restaurants	55	76–77	38
27	Transport services	56–59	78–82	39–42
28	Communication services	60	83	43
29	Banking and financial services	61	84–86	44–47
30	Public services	63	96	52
31	Other services	62, 64	87–95, 97–98	48–51, 53–59

All calculations are done at current producers' prices. It can be safely assumed that the years of the I-O tables mirror the macroeconomic and trade policies in different periods. The year 1979 represents the end of the import substitution period. The year 1985 mirrors export expansion, while the year 1990 reflects import liberalization, 1996 symbolizes instability periods. Finally, the year 2002 represents the policies that follow the restructuring after the financial crisis.

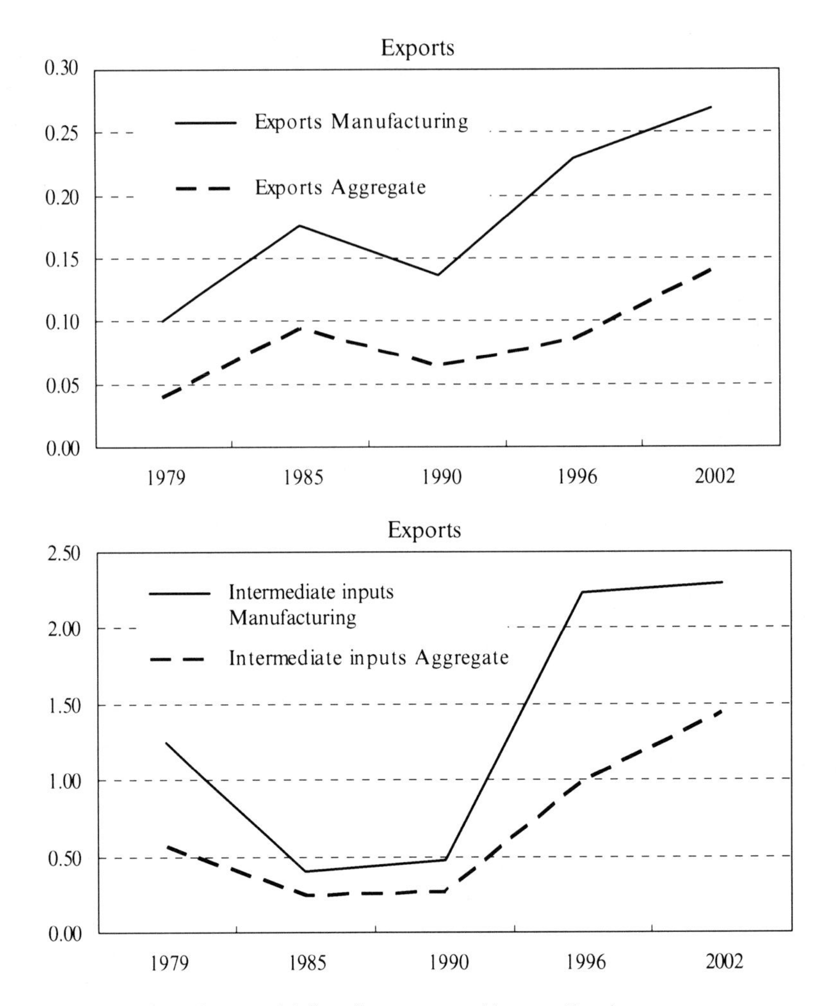

Fig. 9.1 Import dependence multipliers for exports and intermediate inputs

Findings

The main findings are reported in Fig. 9.1 for both the aggregate economy level and the manufacturing sector. Import dependence for both exports and intermediate inputs is seen to have increased from import substitution (1979) to export expansion (1985), as expected. Under the export-led growth strategy (1985–2002), import dependence of both exports and intermediate inputs increased, except for a small decline from 1990 to 1996. From 1979 to 1985 import dependence decreases for intermediate inputs, while it increases for exports. That is, during the transition from import substitution to export promotion, there was import substitution for intermediate inputs but the reverse holds for exports.

Table 9.2 Import dependence multipliers for exports and intermediate inputs by sectors

	Import dependence multipliers for exports					Import dependence multipliers for intermediate inputs				
Number	1979	1985	1990	1996	2002	1979	1985	1990	1996	2002
Raw material-intensive products										
01	0.016	0.022	0.023	0.141	0.149	0.604	0.439	0.550	2.975	3.442
02	0.010	0.056	0.034	0.048	0.032	0.139	0.502	1.063	14.603	3.165
03	0.006	–	–	0.001	–	0.066	0.001	0.004	0.067	0.000
04	0.027	0.545	0.312	0.546	0.499	0.149	1.565	2.103	15.32	1477.0
05	0.004	0.059	0.027	0.045	0.095	0.579	0.254	0.127	1.838	1.017
Labor-intensive products										
06	0.109	–	0.221	0.056	0.101	0.858	0.000	1.085	0.354	0.550
07	0.036	0.046	0.005	0.027	0.014	0.850	0.105	0.013	0.057	0.041
08	0.225	0.015	0.027	0.038	0.024	52.664	0.068	0.907	1.612	0.466
19	0.034	–	0.006	–	0.012	43.814	0.000	0.153	0.000	0.061
10	0.007	0.094	0.083	0.166	0.226	0.007	0.684	0.936	4.966	4.053
11	0.011	0.004	–	–	0.003	4.949	0.076	0.001	0.000	0.194
12	0.059	0.039	0.011	0.008	–	0.601	0.219	0.194	0.099	–
Capital-intensive products										
13	0.148	0.028	0.018	0.056	0.048	1.895	0.174	0.107	0.514	0.936
14	0.013	–	–	0.006	0.001	0.145	0.000	0.000	0.058	0.019
15	0.001	0.021	0.036	0.026	0.086	0.172	0.129	0.454	0.308	0.639
16	0.043	0.017	0.018	0.014	0.027	0.248	0.081	0.051	0.129	0.161
17	0.003	0.491	0.452	0.634	0.683	0.003	0.485	0.810	4.994	3.815
18	0.069	0.019	0.026	0.008	0.055	7.818	0.137	0.103	0.156	0.413
19	0.061	0.048	0.012	–	0.007	4.901	9.344	0.296	–	6.235
Easily imitable R&D-intensive products										
20	0.068	0.342	0.275	0.656	0.809	2.821	2.408	2.170	7.820	24.17
21	0.006	0.049	0.070	0.124	0.146	0.902	0.589	1.296	2.651	1.246
Difficultly imitable R&D-intensive products										
22	0.008	0.044	0.125	0.125	0.251	0.866	0.477	1.586	1.241	1.508
23	0.020	0.074	0.038	0.086	0.161	0.794	1.098	0.799	1.038	0.553
Services										
24	0.018	–	–	–	–	–	–	–	–	–
25	0.003	–	–	–	–	0.049	–	–	–	–
26	0.012	–	–	–	–	1.395	–	–	–	–
27	0.011	–	–	–	0.042	0.154	–	–	–	0.386
28	0.005	–	–	–	0.002	0.132	–	–	0.003	0.154
29	0.003	–	–	0.011	0.030	0.005	–	–	0.388	1.123
30	–	–	–	–	–	–	–	–	–	–
31	0.005	–	–	0.013	–	0.595	–	–	0.032	–

The sectoral composition of import dependence may include important information. Import dependence multipliers for exports (ID^E) are presented in Table 9.2. ID^E reflects increasing dependence on imports or declining import substitution. Intermediate input imports for most or all service sectors are reported as zero in the input–output tables for the years 1985, 1990, and 1996. Therefore, import dependence could not be computed for these sectors. The results for import dependence demonstrate large variations across sectors. During import substitution, ID^E decreased for all sectors except the mining, wood, and chemicals sectors. Under the export-oriented strategy, from 1979 to 2002, ID^E decreased in 9 of the 18 manufacturing industries (food, tobacco, textiles, wearing, wood, furniture, printing, nonmetallic minerals, and fabricated metal products) and increased in the other 9 industries. Notice the general tendency for ID^E to decline from 1985 to 1990 and to increase from 1990 to 2002. To sum up, import dependence of heavy industries such as metals, machinery, equipment, and chemicals increased during the export-oriented era.

The evolution of the dependence of industries to imports for intermediate input use is captured by import dependence multipliers for intermediate input demand (ID^N), which are presented in Table 9.2. The decline in ID^N implies declining import substitution or increasing dependence on imported intermediate inputs. ID^N dropped sharply from 1979 to 1985 for both the aggregate economy and the manufacturing sector as a whole. At the disaggregated level, ID^N dropped in 5 out of the 24 sectors. In all manufacturing industries other than basic metals and transport equipment, ID^N first decreased from 1979 to 1985 and then increased in 2002 relative to the 1985 levels. Overall, from 1979 to 2002, i.e., during the period of export orientation, ID^N dropped in 15 sectors. All these findings imply that import dependence in manufacturing activities for intermediate input use first declined until 1985 but thereafter rose gradually under the export-oriented growth policy.

Therefore, the findings indicate that when the long-term development strategy changed from import substitution type of industrialization to an export-oriented type after 1980, the economy became more dependent on imports for both exports and intermediate inputs. These findings confirm the empirical findings from other related studies (Gunluk-Senesen and Senesen 2001; Senesen and Gunluk-Senesen 2003a, b), which found that the traditional export industries induced more imported inputs after 1980 and that this is in sharp contrast with the aims of the export-oriented policies. The mechanism that led to this outcome is explained by Esiyok (2008) Thus, during the structural adjustment and economic liberalization era after 1980, public companies providing intermediate inputs for the national industries left the stage; as a consequence, the economy became more dependent on intermediate inputs. However, the real appreciation of the domestic currency during this period led to the relative cheapening of imports. Also, due to lack of investments in industries substituting imports necessary for export sectors, the dependence of exports on imported inputs increased.

Development policies during the 1980s and 1990s emphasized market liberalization, encouragement of labor-intensive export activities, and abolishment of quantitative restrictions on imports. The structural changes in the economy resulted in changes to comparative advantage. As such, some exports increased, and some imports (particularly intermediate inputs) did as well. However, a major consequence of import liberalization is an increase in the dependence of the economy, export industries in particular, on imports especially of intermediate inputs.

Why did import liberalization and export promotion strategy yield mixed and partially unfavorable results regarding the import dependence of the economy? The answer to this question may be found from findings of related other studies which have assessed the development strategies. Pamukcu and de Boer (2000) found for the period 1979–1990 that trade liberalization brought about an increase in import substitution of final goods and intermediate materials by local production, which, in turn, led to increases in total and manufacturing imports. The contribution of export expansion to investment good imports was small, indicating that imports resulting from export expansion were of low skill and technology intensive, Guncavdi et al. (2008) confirmed that while terms of trade have deteriorated after 1980, import dependence for exports increased largely. Worsening terms of trade with increasing dependence on imported inputs reduces the income generation capacity of exports.

In addition to I-O studies, econometric studies also analyzed the export-oriented strategy and the increasing openness in the Turkish economy (i.e., trade and capital liberalization) during the 1980s and 1990s. Yeldan et al. (2000) found that the increasing openness in the Turkish economy during the post-1980 era did not result in enhanced competition, but rather led to increasing profit margins in the manufacturing industries. They also argue that the government provided these industries with implicit protection through tax exemptions, and rent transfers in the form of privatizations. Olgun and Togan (1991) have also shown that export industries were more protected than the industries that compete with imports. These observations lend support to the argument that industrialization, which was adopted as a national goal during the 1960s and 1970s, was neglected during the 1980s and 1990s (Senses and Taymaz 2003).

Implications for Technological Content of Exports, Technology Policies, and Technology Transfer

The following two subsections draw some policy implications from the empirical findings of import dependence for technology policies, knowledge-oriented industrial policies, and technology transfer. Turkey has adopted an outward-oriented and export-oriented industrialization strategy. Therefore, policy recommendations should emphasize competitiveness of Turkey's exports, in conjunction with various issues related to technology transfer.

Exports, Intermediate Inputs, Import Dependence, and Technological Content

The results in Table 9.2 show that during the outward-oriented era after 1980, import dependence for both easily imitable and difficulty imitable R&D-intensive

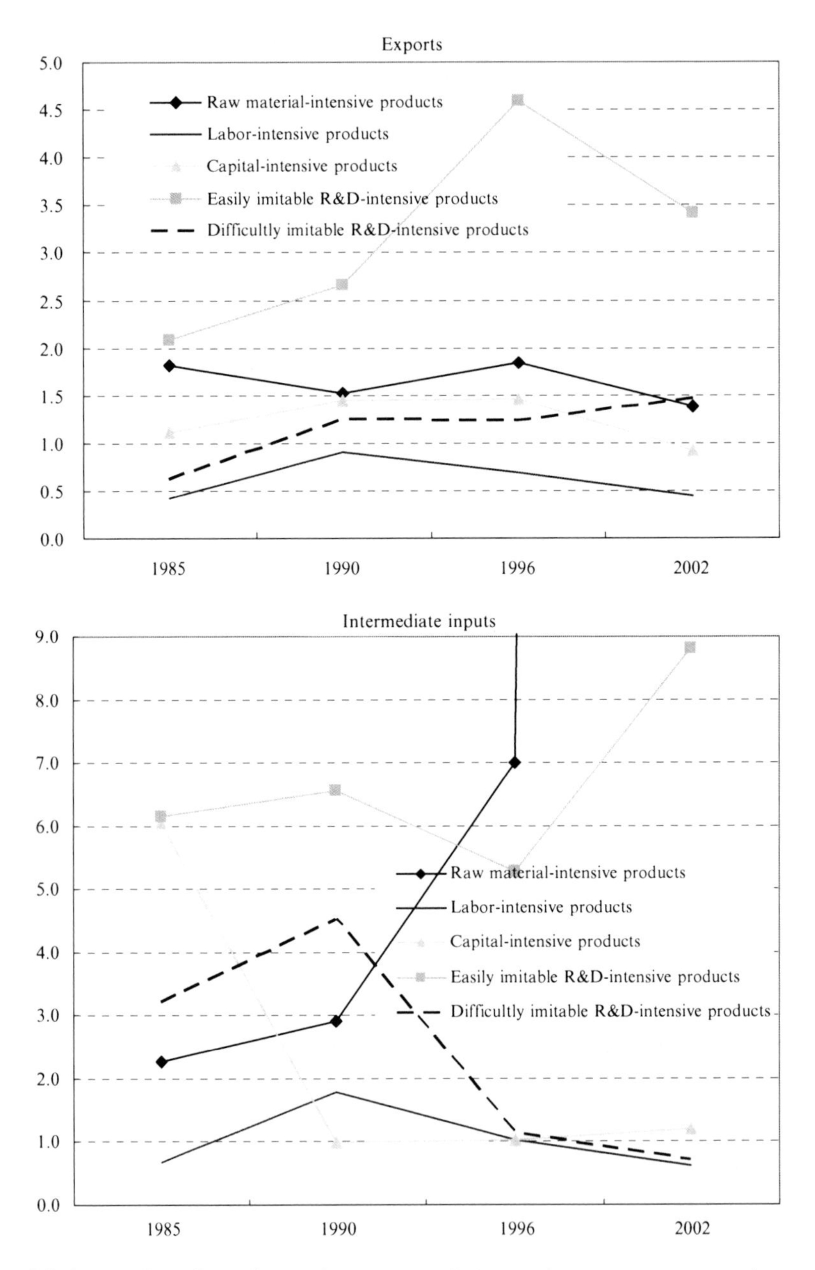

Fig. 9.2 Import dependence by product groups relative to the economy average (economy average = 1.0)

manufactured exports increased over time. Sectors are aggregated into broader product groups in Fig. 9.2. Import dependence for both exports and intermediate inputs is presented for these broader product groups, excluding services. The figures in the table are reported relative to the economy average. Of importance here are the

R&D-intensive products, both easily imitable and difficultly imitable. It is observed in Fig. 9.2 that exports of easily imitable R&D-intensive products are the highest in the import-dependent sector, while the difficulty imitable R&D-intensive products are the second highest in 2002. In the case of intermediate input usage, easily imitable R&D-intensive products are the second highest next to the raw material-intensive products. It should be noted that Turkey is extremely dependent on raw materials such as oil. On the other hand, as with other major product groups, the dependence of difficultly imitable R&D-intensive products has decreased steadily after 1990 and below the economy average in 2002.

These findings have important implications. The share of easily imitable R&D-intensive products in total exports and total imports is shown in Fig. 9.3. Services exports and imports are excluded from the totals. The share of difficultly imitable R&D-intensive exports has risen sharply after 1990. Although total R&D-intensive exports made up only about one-fourth of total exports in 2002, this is a big improvement since 1985 (slightly over one-tenth). In the case of imports, R&D-intensive products, easily and difficultly imitable combined together, made up about four-tenths of total imports in 2002, up from about one-fifth in 1985.

As a result, R&D intensity has been rising since 1985 but imports have risen more than exports. The figures shown in Fig. 9.3 are compiled from the I-O tables, with 2002 being the most recent year. Using the official trade data published by the Undersecretariat of Foreign Trade, Fig. 9.4 reports the shares of the stated products in 2009. The situation for imports has not changed much since 2002. However, dependence on difficultly imitable R&D-intensive products has increased significantly to about one-fourth of total imports, while the dependence of easily imitable products has decreased

All the above findings show that the import dependence of the Turkish economy for exports and intermediate inputs has increased since the 1980s and has increased more rapidly in recent years. In addition, the share of R&D-intensive products in these imports has also risen. That is to say, the economy has experienced an increase in the technological content of its exports to some extent, but that of the imports necessary to produce the products destined to both domestic and foreign markets has risen more, Turkey has become more dependent on foreign technology. These findings have significant policy implications for Turkey. The development strategy adopted by the current government is outward-oriented industrialization with a special emphasis on labor-intensive industries. Thus, sustaining the competitiveness of national export industries with their increasing dependence on imports needs to be evaluated carefully.

In this regard, the most noticeable challenge in the recent decade is the competition with emerging market-originated exports, most notably China and India. With the rise of these two latter countries in world markets, there has been a growing concern among developing countries about competing with these two countries, especially after China's WTO accession in late 2001 (Eichengreen et al. 2007; Greenaway et al. 2008). The structure of Chinese exports has been continuously changing from light manufactures to more capital-intensive machinery and equipment. China's factor endowment is largely similar to that of Turkey. In the OECD

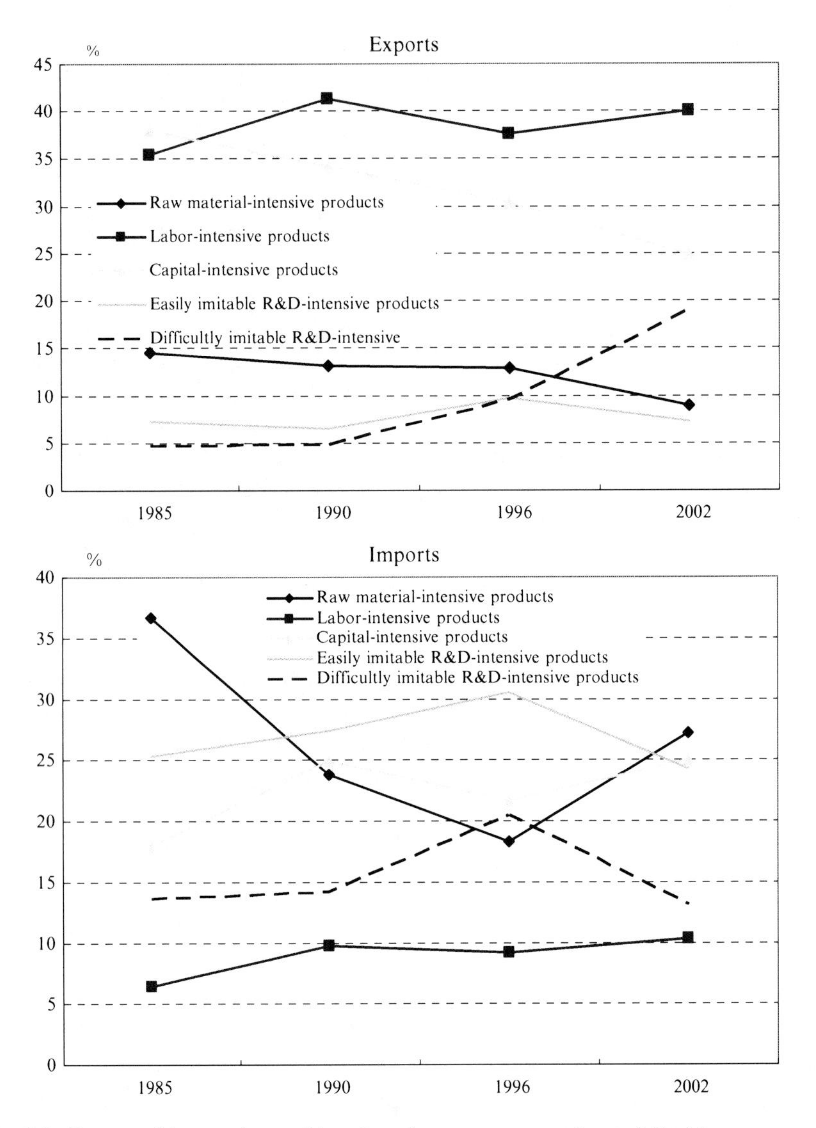

Fig. 9.3 Export and import shares of broad product groups, according to I-O tables

market, Turkey has comparative advantages in labor-intensive and raw-materials products, while China has a comparative advantage in labor-intensive and easily imitable products. According to Seymen and Simsek (2006) who used the revealed comparative advantages and marginal intra-industry trade indices for the period 1995–2002 both countries have comparative advantages in more or less similar products. This poses a threat for Turkey because China has a cost advantage in the resources that are used intensively in the making of these products. On the other hand, it is noteworthy that Akkemik and Goksal (2010) found that Turkish exports

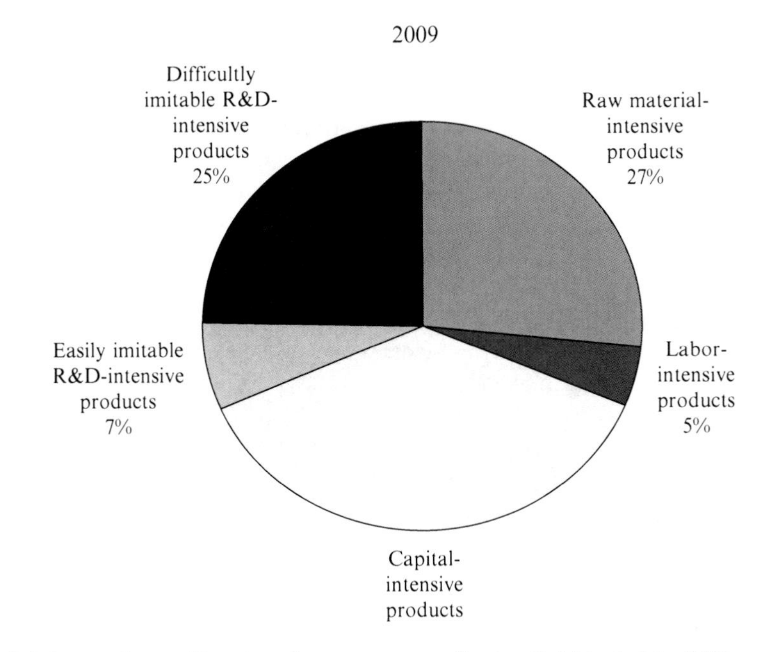

Fig. 9.4 Import shares of broad product groups, according to official trade data, 2009

are not affected by the real effective exchange rate (i.e., price competition with China), and that competitiveness can be explained only by real factors that impact on productivity and technology. Related to this finding, Atasoy and Saxena (2006) reported that during the period, for which the Turkish Lira has been floating (after 2001), the actual real effective exchange rate was close to the equilibrium level.

Having established the empirical facts that the dependence of Turkish products has increased, the competitiveness of Turkish exports relies more on "real" factors that impact on technological capability rather than price competitiveness.

Issues Related to Science and Technology Policies and Internalization of R&D

As a general theoretical argument for traditional industrial policies, it is considered that due to market failures and public good nature of knowledge, private sector R&D falls short of socially optimal level. This argument calls for the government to play an active role in the area of R&D. In the era of new economy, this intervention takes a different form, as compared to "picking the winners" in the traditional industrial policy. Akkemik (2009) reviews traditional industrial policies in the East Asian context. In the knowledge-oriented industrial policy, the linkages and interactions between the stakeholders are addressed and the ways to strengthen these relations are sought after Dobrinsky (2009).

The empirical findings above indicate that Turkey has become increasingly dependent on foreign technology. This should also be assessed with the increasing importance of enhancing competitiveness in the era of the "new economy." In this regard, it becomes a prerequisite for a government to design and effectively implement knowledge-oriented industrial policies, which will support and facilitate generation, accumulation, and dissemination of knowledge (Dobrinsky 2009). This type of industrial policy is very different from the traditional industrial policy. The aim of the traditional policy is to nurture selected "winner" industries. The selection is based on predetermined goals such as achieving scale economies, enhancing social welfare through employment creation, and raising value-added content.

As far as the acquisition of these technologies is concerned, policies related to the transfer of foreign technologies through worldwide internationalization of R&D have gained importance.

Deepening relations between emerging economies, including the BRICS (Brazil, Russia, China, India, and South Africa), and the Triad countries (US, Japan, and the EU) have improved cooperation and competition in science and technology policies. On the other hand, China and India have emerged as new sources of innovation and technology. This situation creates more opportunities for developing countries like Turkey in the areas of technology policy cooperation and technology transfer other than the Triad. However, in terms of competition for FDI from advanced economies of the Triad, these countries also bring up challenges against Turkey. In this regard, internalizing the benefits of the internationalization of R&D requires well-designed science and technology policies and knowledge-oriented industrial policies. Therefore, priority areas, objectives, policy instruments, relevant strategies, and the roles of the stakeholders (public and private) should be defined and determined carefully and clearly.

Conclusion: Technology Transfer and Policies to Encourage Innovation

Most researchers tend to blame the post-1980 industrialization strategy for increasing dependence of domestic industries as well as export industries on such imports. An important pillar of the development policies since the early 1980s was import liberalization, which is found to result in increasing dependence of the economy on imports, especially of intermediate goods. We showed above that Turkish industry, export industries in particular, has become increasingly dependent on imports, especially those with difficult-to-imitate technologies.

The increase in the import dependence of exports points to an important shortcoming in the Turkish economy. However, devising a wise technology transfer policy can remedy this important problem. It has long been discussed among economists whether government support for R&D is a substitute or a complement for private sector R&D activities. In the new economy, the latter seems to be vitally important for the Turkish economy. There are strong reasons, based on the results of the empirical findings of this study, to believe that a well-designed knowledge-oriented industrial policy will have a favorable impact on national welfare in Turkey in the near future.

References

Akkemik KA (2009) Industrial development in East Asia: a comparative look at Japan, Korea, Taiwan, and Singapore. World Scientific, Hackensack

Akkemik KA, Goksal K (2010) Do Chinese exports crowd-out Turkish exports? Iktisat Isletme ve Finans 25(287):9–32

Atasoy D, Saxena SC (2006) Misaligned? Overvalued? The untold story of the Turkish Lira. Emerg Market Finance Trade 42(3):29–45

Boratav K, Turel O, Yeldan E (1996) Dilemmas of structural adjustment and environmental policies under instability: post-1980 Turkey. World Dev 24:373–393

Boratav K, Yeldan E, Kose A (2000) Globalization, distribution, and social policy; Turkey: 1980-1998. CEPA and New School for Social Research, Working Paper No. 20

Bulmer-Thomas V (1982) Input-output analysis in developing countries. Wiley, New York

Dobrinsky R (2009) The paradigm of knowledge-oriented industrial policy. J Ind Compet Trade 9:273–305

Edwards S (1998) Openness, productivity and growth: what do we really know? Econ J 108:383–398

Eichengreen B, Rhee Y, Tong H (2007) China and the exports of other Asian countries. Rev World Econ 143(2):201–226

Esiyok A (2008) Turkiye Ekonomisinde Uretimin ve Ihracatin Ithalata Bagimliligi, Dis Ticaretin Yapisi: Girdi-Cikti Modeline Dayali Bir Analiz (Import dependence of production and exports in Turkey, the structure of foreign trade: an analisis based on the input-output model). Uluslararasi Ekonomi ve Dis Ticaret Politikalari 3(1–2):117–160

Greenaway D, Mahabir A, Milner C (2008) Has China displaced other Asian countries' exports? China Econ Rev 19:152–169

Guncavdi O, Kucukcifci S, Ungor M (2008) Cari Aciklar ve Turkiye Ekonomisinin Artan Doviz Ihtiyaci (Current account deficits and the rising foreign exchange demand of the Turkish economy). Uluslararasi Ekonomi ve Dis Ticaret Politikalari 3(1–2):57–84

Gunluk-Senesen G, Senesen U (2001) Reconsidering import dependency in Turkey: the breakdown of sectoral demands with respect to suppliers. Econ Syst Res 13(4):417–428

He D, Zhang W (2010) How dependent is the Chinese economy on exports and in what sense has its growth been export-led? J Asian Econ 21:87–104

Kim W (1990) A comparative analysis of export-led growth: case study of South Korea and Taiwan. Unpublished PhD Dissertation, Catholic University of America

Olgun H, Togan S (1991) Trade liberalization and the structure of protection in Turkey in the 1980s: a quantitative analysis. Rev World Econ 127(1):152–170

Pamukcu T, de Boer P (2000) Determinants of imports of Turkey: an application of structural decomposition analysis – 1968-1990. Yapi Kredi Econ Rev 11:3–27

Sachs JD, Warner AM (1995) Economic reform and the process of global integration. Brookings Pap Econ Act 1:1–118

Senesen U, Gunluk-Senesen G (2003a) Import dependency of production in turkey: structural change from 1970s to 1990s. Paper presented at 10th annual conference of the economic research forum, Cairo

Senesen U, Gunluk-Senesen G (2003b) Uretimde Disalima Bagimlilik: 1970'lerden 2000'lere Ne Degisti? (Import dependency of production: what has changed from 1970's to 2000's?). In:

Kose AH, Senses F, Yeldan E (eds) Iktisat Uzerine Yazilar II: Iktisadi Kalkinma, Kriz ve Istikrar (Papers on economics II: economic development, crisis and stability). Iletisim Yayinlari, Istanbul, pp 533–559

Senses F, Taymaz E (2003) Unutulan Bir Toplumsal Amac: Sanayilesme Ne Oluyor? Ne Olmali? (A forgotten social objective: what is happening with industrialization? What should?). In: Kose AH, Senses F, Yeldan E (eds) Iktisat Uzerine Yazilar II: Iktisadi Kalkinma, Kriz ve Istikrar (Papers on economics II: economic development, crisis and stability). Iletisim Yayinlari, Istanbul, pp 429–461

Seymen D, Simsek N (2006) Turkiye ile Cin'in OECD Pazarinda Rekabet Gucu Karsilastirmasi (Comparison of competitive powers of Turkey and China in the OECD market). Iktisat Isletme ve Finans 21(244):38–50

Yeldan E (1989) Structural adjustment and trade in Turkey: investigating the alternatives beyond export-led growth. J Pol Model 11:273–296

Yeldan E, Ozcan KM, Voyvoda E (2000) 1980-Sonrasi Türk Imalat Sanayinin Dinamikleri Uzerine Gozlemler (Observations on the post-1980 dynamics of the Turkish manufacturing industry). Iktisat Isletme ve Finans 15(176):22–37

Chapter 10
Tertiary Education and R&D in Turkey: An Assessment and Policy Implications

Ahmet Kesik

Introduction

This study aims to examine the current situation of tertiary education in Turkey within the global competitive environment and undertake an analysis of the resources allocated for instruction and research. Performance data are also presented.

In recent years, Turkey has established many public universities as well as those run by private foundations. Furthermore, new public universities are likely to be opened through international partnerships in the near future. Such increases in numbers would also raise the supply of undergraduate education significantly. In response, the Turkish tertiary education sector should also adapt itself not only to the shift in the supply-demand balance, but also to the growing international and regional competition in this particular area.

In the light of these developments, it becomes crucial how public universities that continue to remain bound with bureaucratic rules in tertiary education and research will redefine their missions in order to develop their sustainable competitive strategies. Thus, two aspects are studied here: (a) diversity of the public universities and how they will reflect this diversity in their missions and (b) how the public universities will develop and implement a sustainable competitive strategy within the framework drawn by the old bureaucratic rules that limit the comprehension of the recent developments. This would mean that public preferences and policies on the tertiary education would need to be fine-tuned to take these institutions to higher levels of quality within the global context.

The share of the resources used by universities in Turkey is seen to increase within both public sector and private sector expenditures. The outstanding rise in the number of universities and students, on the one hand, and the need for efficient

A. Kesik (✉)
Ministry of Finance, Ankara, Turkey
e-mail: ahmet.kesik@sgb.gov.tr

M.A. Yülek and T.K. Taylor (eds.), *Designing Public Procurement Policy in Developing Countries*, DOI 10.1007/978-1-4614-1442-1_10,

and effective use of these resources to enable service delivery at higher levels of service quality in tertiary education in the future, on the other, have become important issues for both the public and the household. There has been an increase in the number of universities providing tertiary education service from 70 in the year 2000 to 154 in 2010, and a rise in the number of students from 1.1 to 1.8 million during this period.

The issue of state intervention in the education system, the resources allocated for instruction and research in comparison with the international standards and the performance results will all be taken up within the present study.

A Conceptual Framework for Tertiary Education

Economic development is becoming increasingly associated with the capacity of a nation to acquire and use knowledge. For example, although Ghana and Korea were at the same point in terms of their real GDP per capita in 1960, Korea outscored Ghana by raising its real GDP per capita to a level that was almost 3.5 times that of Ghana in the period 1960–2005 (Fig. 10.1). The fundamental reason behind this great leap was no doubt the total factor productivity growth or knowledge accumulation in Korea as compared to Ghana.

As seen from this particular example, acquisition and use of knowledge is one of the major sources for the postindustrial countries in reaching higher levels of prosperity. In this regard, it should be noted that tertiary education makes an undeniable contribution to the creation of human capital which can use information technologies and transfer them into production.

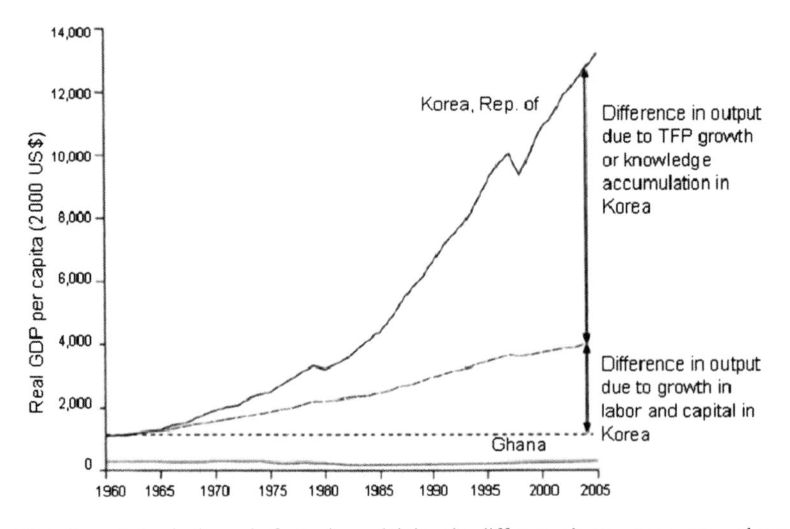

Fig. 10.1 Knowledge is the main factor in explaining the difference between poverty and wealth. *Source*: WBI (2007), p. 4

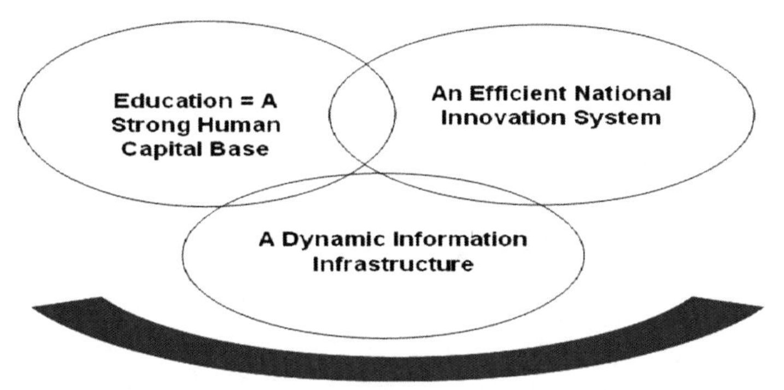

Fig. 10.2 Key strategic dimensions for transition to a knowledge-based economy. *Source*: Salmi (2009), p. 2

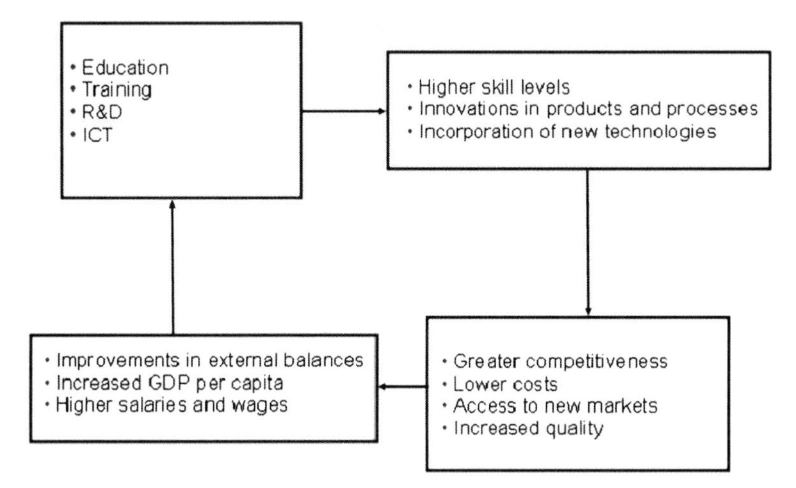

Fig. 10.3 Knowledge-based competitiveness and growth. *Source*: WBI (2007), p. 28

Four basic strategies are envisaged for the transformation to a knowledge-based economy (Salmi 2009, p. 4), namely a sound human capital infrastructure, an efficient national innovation structure, a dynamic knowledge infrastructure, and an appropriate economic and institutional structure.

As seen in Fig. 10.2, when countries develop in terms of education, training, R&D, and ICT (Information and Communications Technology) sectors, then these countries will be able to create higher skill levels, innovations in products and processes, and achieve incorporation of new technologies. These achievements, in turn, will result in greater competitiveness, lower costs, and access to new markets and increased quality. As a consequence, this will lead to improvements in external balances, increased per capita, and higher salaries and wages. Again, these outcomes will allow countries to undertake more expenditure for knowledge accumulation through education, training, R&D, and ICT. This will continue as a closed loop (Fig. 10.3).

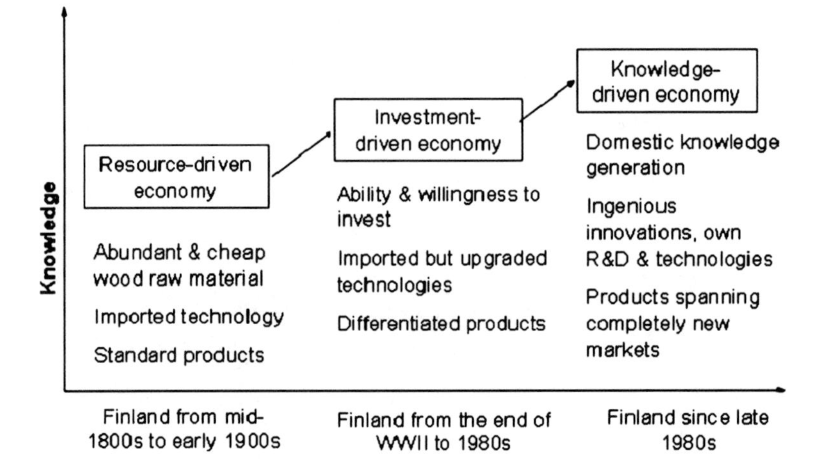

Fig. 10.4 The stages of industrial and economic development in Finland. *Source*: WBI (2007), p. 49

As a matter of fact, when looking at the industrial and economic development stages of Finland (Fig. 10.4), a resource-based economy (cheap and abundant wood as raw materials, imported technology, and traditional products) can be seen till the early 1900s; an investment-based economy (ability and willingness to invest, imported but upgraded technology and diversified products) beginning from the World War II to the 1980s; and a knowledge-based economy (domestic knowledge production, innovation, domestic R&D technologies, and a range of products totally introduced to new markets) since the 1980s (WBI 2007, p. 49).

Thus, it is not only that acquisition and use of knowledge is a major factor for the postindustrial countries to reach prosperity, but it also brings about greater social benefits than social costs. Furthermore, the prospect of increasing efficiency in the economy by developing human capital places the educational expenditures to a special niche among other sectoral expenditures. Moreover, the governments choose to intervene in this sector because a considerable part of the educational expenditures are considered to ensuring social equity, provide equal opportunities, and carry high positive externalities (Yilmaz 2007, p. 43).

Generally accepted grounds for the governmental interventions in the tertiary education can be listed as follows:

(a) Delivery of tertiary education service to all sectors of the society
(b) External benefits of the tertiary education, inability to divide them and to reflect them to the prices
(c) Need for strengthening the democratic structure
(d) Protection of the low-income individuals' right to be educated at the tertiary level
(e) Assumption of the existence of a relationship between the tertiary education and economic growth

It is possible to argue that tertiary education has both direct and indirect influences on an economy. It has direct influence on the economy because the qualified labor force is among the production factors. On the other hand, it has an indirect influence on the leading strategic factors of economic development through increases in productivity and investment capacity of an economy (Korkmaz 1975, pp. 133–134). In this respect, it is possible to analyze the influence of tertiary education on economic development under the titles of an increase in productivity, technology, efficient resource allocation, and social development.

Therefore, capital stock and physical capital capacity, as well as knowledge and capacity for implementation of this knowledge efficiently, are important variables in the definition of the economic growth. In this respect, technological development that is created as a result of investments in the human capital and the use of new technologies through human capital stock will enable higher level production with few additions to the physical capital stock. However, it should also be noted that making investments in human capital without making sufficient investments in physical capital may lead to unemployment and lower economic growth. Therefore, it should be emphasized that physical and human capital investments are complementary rather than being alternatives to one another.

Contrary to popular belief, the relationship between educational expenditures and growth may not be so clear all the time. This is mainly because educational expenditures are not included in the financial statistics in a detailed way or, more importantly, there is a long-term or lagged relationship between expenditures for education and participation in the labor force (Emil and Yilmaz 2003, p. 9).

Additionally, although it is generally desired that the governments take part in the supply and financing of the education services, there is not a consensus on which role should be undertaken by the government and to what extent. However, governments have taken part in delivering tertiary education service in many developing and developed countries for a long time. Expenditures for the tertiary education still have an important share in the state budgets, although to a lesser extent as compared to the past.

Tertiary education serves the following three main functions:

(a) Instruction: It is the most basic activity and has a social benefit.
(b) Scientific research: One mission of tertiary education is to expand and spread knowledge.
(c) Public service: There is a service relationship between the tertiary education institutions and the society. These are cultural, social and sports activities, and certain services rendered to the public, such as healthcare services.

Tertiary education institutions may operate within the scope of one or all of these functions. However, it should be noted that today these functions are engaged in one another and it is very difficult to make certain differentiations among them. For example, an activity carried out within the scope of the scientific research function may also serve the function of public service. On the other hand, a scientific research function may also include the function of instruction in some cases. What is important here is the efficient and result-oriented performance of these functions within the scope of the tertiary education service and the value-addition created in the economy and within the production processes.

Educational Expenditures and Universities from a Comparative Point of View

The Ratio of the total educational expenditures to the GDP ranges from 5 to 6% in many countries today. The average rate in the OECD is 6%. In some countries in the OECD the average reaches 8%, while in others it falls to 6%. Looking at the OECD countries in general, for countries exceeding the average rate (excluding countries such as Finland, Sweden, Belgium, etc.), the primary factor affecting this is expenditures made by the nonpublic actors.

The OECD average share of educational expenditures by the nonpublic sector in the total educational expenditures came to 13.8% in 2006, while this same rate was 9.1% above the average for 19 EU member states. The share of the private sector varies significantly among the countries. The share of the nonpublic sector decreases to 1–3% in the countries such as Finland, Sweden, etc., while it rises above 30% in the USA, Korea, Japan, etc.

Turkey is below the OECD average in terms of the ratio of educational expenditures to the GDP (Fig. 10.5). The figure for the OECD as a whole was 5.9% in 2002, while it was 5.4% in Turkey (TURKSTAT 2002; Yilmaz 2007, p. 189). The 2006 data show that the OECD average for educational expenditures reaches 5.8%, while it ranges from 4.5 to 5% for Turkey in the light of estimates within the scope of this study.

Although Turkey is approaching the OECD level in terms of educational expenditures, an outstanding difference is seen when the public and private expenditures are viewed separately in Turkey (especially household private spending). Looking at the public expenditures to GDP ratio, there is still a significant difference in the levels of the OECD and Turkey. As a matter of fact, the public

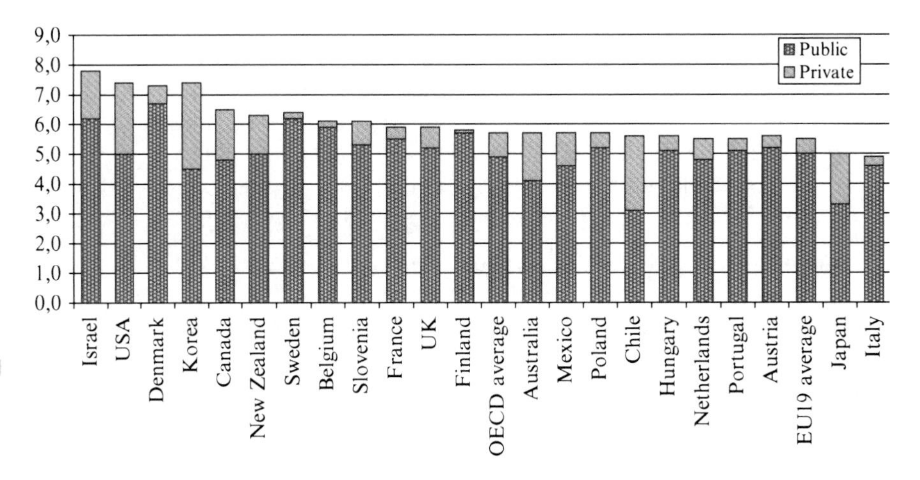

Fig. 10.5 Total educational expenditures/GDP ratios of the countries (2006, %). *Source*: OECD (2009)

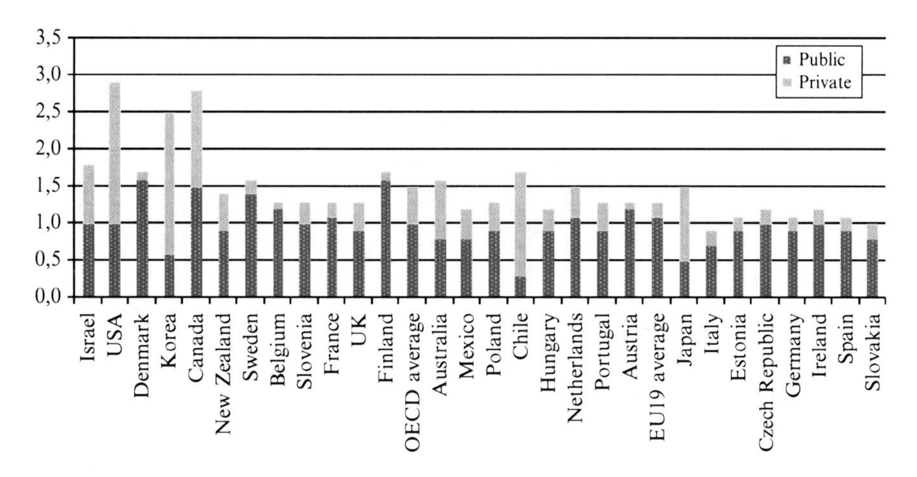

Fig. 10.6 Ratio of expenditures on universities to GDP by countries (2006, %). *Source*: OECD (2009)

expenditures to GDP ratio of the OECD was 3.4% in 2006, whereas this ratio was just 1.9% in the data for Turkey (we estimate that this rate cannot be calculated accurately because of the reasons resulting from data standardization for Turkey, excluding those for the tertiary education and also because of the calculation method of the OECD data). Furthermore, it is seen that public resources allocated by Turkey for all educational levels in total correspond to 55% of the OECD average.

Along with the overall figures, when public resources are considered for each level of education, the difference from the OECD average becomes clearer. For example, Turkey approaches the OECD average by 80% in tertiary education, whereas this ratio remains at 55% in the primary and secondary education in total.

Today, almost one-fourth of the total educational expenditures are made for the purpose of tertiary education, while the remaining 75% is used in other levels of education. The average ratio of public expenditures to the GDP is 1.5% in the OECD countries, but deviation among the countries becomes higher at levels compared to the total educational expenditures (Fig. 10.6). Countries such as the USA, Korea, and Canada have an expenditure size doubling the average. The share of the private spending in the educational expenditures for the universities is at higher levels, as expected; and it is important to see differences in this rate among the countries in terms of variations in the policies implemented.

There are some differences among the OECD countries in terms of public resources allocated for tertiary education (Fig. 10.7), while the ratio of expenditures to the GDP is 1% of the average. The EU average is above the OECD average, and Turkey ranks near the bottom of the list among the OECD countries in terms of public resources. However, when analyzed separately in terms of different levels of education, Turkey takes its place among the countries allocating relatively more resources to the tertiary education.

The ratio of educational expenditures to the GDP followed a fluctuating course during the period. This ratio was 3.2% in 2000, while it decreased to 3% in 2005

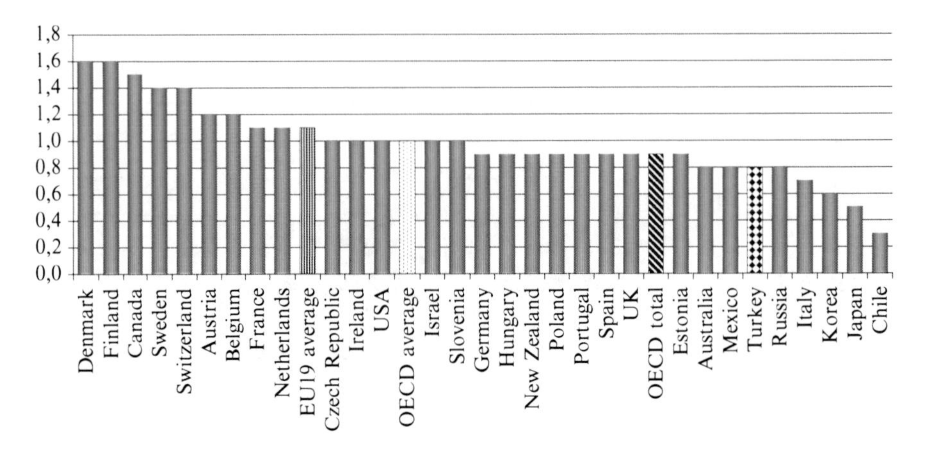

Fig. 10.7 Ratio of public resources allocated for tertiary education to GDP in the OECD countries. *Source*: OECD (2009)

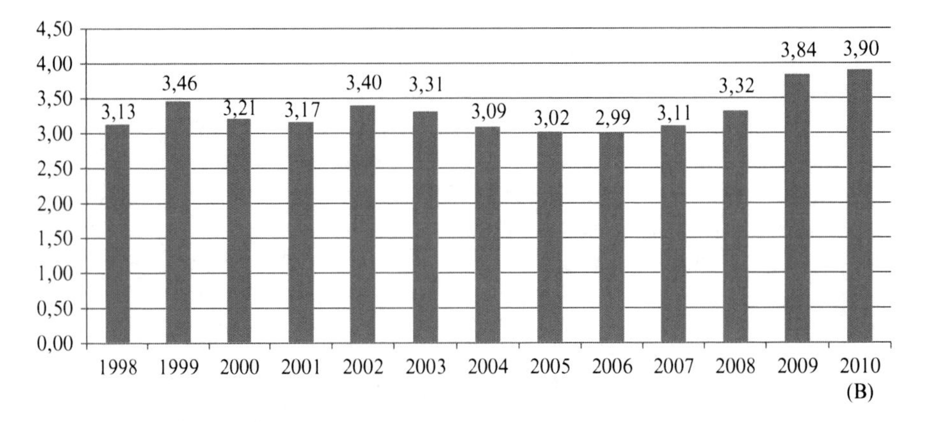

Fig. 10.8 Ratios of educational expenditures to GDP in Turkey by years. *Source*: Yilmaz and Emil (2008), Yilmaz (2007)

and 2006. Beginning to increase after 2008, 2009, and 2010, the seasonal effect of the economic contraction should also be considered (Fig. 10.8).

The ratio of public resources allocated for tertiary education for educational purposes to GDP ranged from 0.8 to 1% in the last 10 years. The share of the educational expenditures used by the universities in the total educational expenditures is 20–22% (educational expenditures include revolving funds and in-kind expenditures) (Fig. 10.9). This figure increases to 27% when the educational expenditures made by the Directorate General of Higher Education Credit and Hostels Institution are also included. Considering this ratio is around 22% in the OECD countries, this figure in Turkey is 5% points above the OECD average in terms of the share of the resources allocated for the tertiary education in the total educational expenditures.

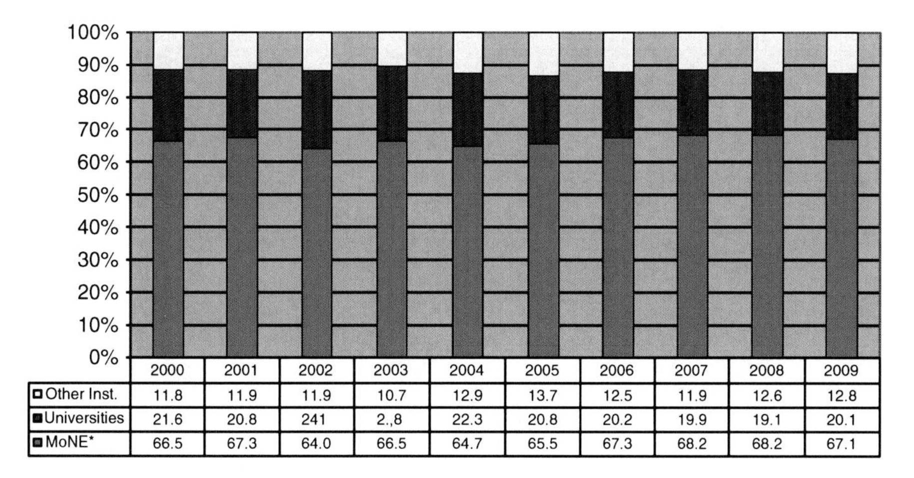

	2000	2001	2002	2003	2004	2005	2006	2007	2008	2009
Other Inst.	11.8	11.9	11.9	10.7	12.9	13.7	12.5	11.9	12.6	12.8
Universities	21.6	20.8	241	2.,8	22.3	20.8	20.2	19.9	19.1	20.1
MoNE*	66.5	67.3	64.0	66.5	64.7	65.5	67.3	68.2	68.2	67.1

Fig. 10.9 Distribution of the public educational expenditures by agencies in Turkey. *Source*: Yilmaz and Emil (2008), Yilmaz (2007). *MoNE: Ministry of National Education

Financial Structure and Development of the Universities in Turkey

Statistics on the Turkish University System

There is an outstanding increase in the number of both public and foundation (private) universities in the 2000s in Turkey. The number of the universities in Turkey has increased from 70 in 2000 to 154 in 2010, an increase of more than 100% in the period 2000–2010 (Table 10.1).

The fastest increase in the number of public universities was experienced in 2006, 2007, and 2010. The rise in the number of foundation universities followed a steady course, the acceleration becoming more prominent after 2007.

The number of academicians and students in the public universities increased considerably in the period 2000–2010, while the number of students has increased by 80% from 1.1 million in 2000 to 1.94 million in 2010. Similarly, the number of the academicians also rose from 24,000 to 41,500 in the same period (Table 10.2; Fig. 10.10).

As seen above, there is not a significant change in the number of students per academicians and total teaching staff. The number of students per academician remained the same at 45 students, while teaching staff exhibited a slight increase from 16 in 2000–2001 to 18 in 2009–2010.

Table 10.1 Number of the universities by years

	Public universities	Foundation universities	Total
2000	52	18	70
2001	53	20	73
2002	53	22	75
2003	53	24	77
2004	53	26	79
2005	53	27	80
2006	68	28	96
2007	85	29	116
2008	94	33	127
2009	94	39	133
2010	102	52	154

Source: The Council of Higher Education (YOK)

Table 10.2 Number of the academicians and students in public universities by years

Years	Number of academics	Total number of teaching and research staff	Number of students	Number of students per academics	Number of students per total teaching staff
2000–2001	23,975	66,720	1,078,879	45	16
2001–2002	25,953	69,987	1,142,114	44	16
2002–2003	27,617	74,104	1,232,255	45	17
2003–2004	29,075	77,042	1,294,172	45	17
2004–2005	30,668	79,533	1,377,837	45	17
2006–2006	32,095	82,220	1,377,837	43	17
2006–2007	34,116	86,487	1,573,803	46	18
2007–2008	37,820	96,075	1,619,501	43	17
2008–2009	38,911	97,891	1,746,534	45	18
2009–2010	41,553	102,647	1,936,602	47	19

Source: Statistics from Student Selection and Placement Center (OSYM 2010) and YOK (August 2010)
Note: Student numbers reflect the total figure, excluding those in Open Education

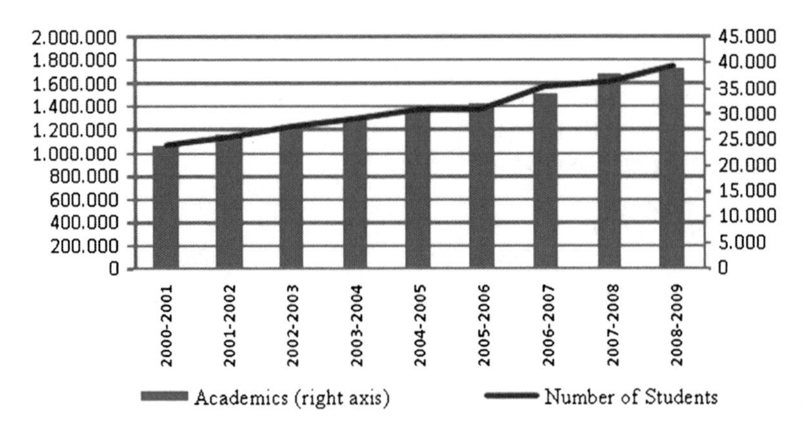

Fig. 10.10 Number of the academicians and students (2000–2009). *Source*: Table 10.2

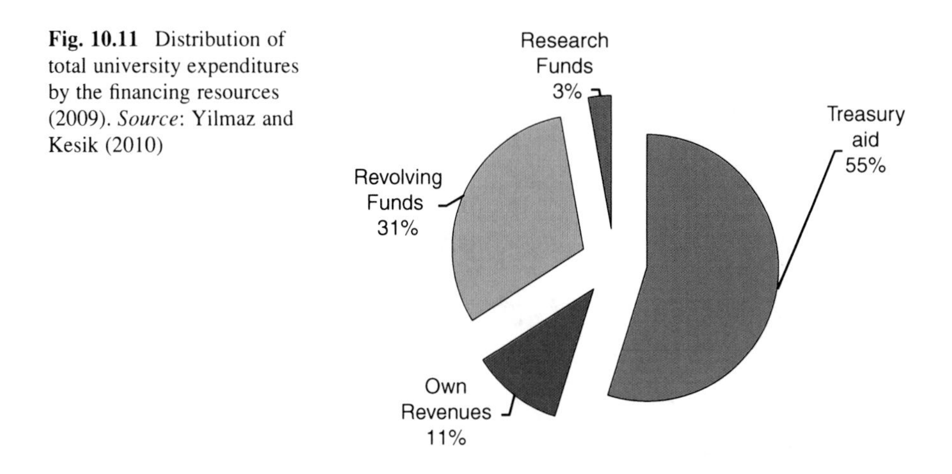

Fig. 10.11 Distribution of total university expenditures by the financing resources (2009). *Source*: Yilmaz and Kesik (2010)

Resources Used by the Public Universities in Turkey

Expenditures by the universities as special budget administrations reached 9 billion TL in total within the scope of central government budget for 2009. When the revolving funds revenue of 4.4 billion TL is added to this figure, total expenditures rise to 13.4 billion TL. Total expenditures as a percentage of the GDP amounted to 1.4 in 2009. In 2010, the budget appropriation was increased by 4.1%, envisaged to reach 9.4 billion TL.

Approximately 55% of the universities' expenditures are composed of treasury support reflecting contributions of the taxpayers, and 11% of them are composed of their own revenues (Fig. 10.11). Within this framework, direct budget resources are ranging between 65 and 70% of the total resources. The third biggest resource of the universities is the revolving funds revenues. The share of the revolving funds revenues has increased to above 30% recently, particularly because of the healthcare service deliveries by the medical school hospitals. This figure approximates 40% in the universities with big hospitals such as Ankara, Hacettepe, and Istanbul Universities. However, this figure decreases considerably in the nonmedical universities. The share of the external project credits, particularly the EU projects, some internal project resources, and research funds transferred from the revolving funds revenues, reached 3% of total resources of the universities (in these figures, research fund cuts considered within the scope of revolving funds expenditures are deducted from the revolving funds amount and reflected to the research revenues). Since the full costing system is not used in the research projects and contracts in Turkey, it can be said that research expenditures are calculated at levels below the real value. Once the full costing system envisaged by the European Union is fully adopted in consistent with the budget, it will be possible to calculate the resources allocated for the research more accurately.

When looking at the budget-related expenditures (Treasury aid) of the universities with fixed prices, it is seen that the expenditures from 10 billion TL

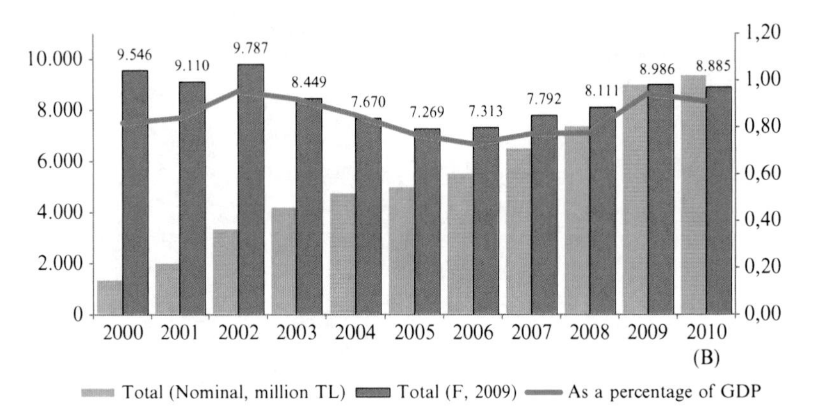

Fig. 10.12 Budget-related expenditures of the universities for 2000–2010. *Source*: Yilmaz and Kesik (2010)

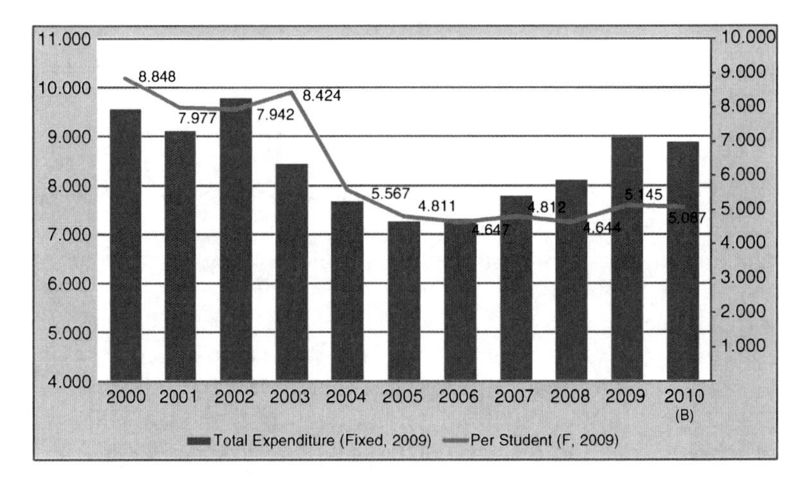

Fig. 10.13 Expenditures per student (with fixed prices, 2009). *Source*: Yilmaz and Kesik (2010)

in 2000 to 7.3 billion TL in 2005 and then rose to 9 billion TL in 2009 by gaining an impetus as of 2007 (Fig. 10.12). As a percentage of the GDP, this rate came to 0.8% in the early 2000 and decreased to 0.7% in the same period and then increased to slightly above 0.9% with the rise in 2008.

Furthermore, the increase in the number of universities and students in this period brings on the need for evaluating university expenditures on the basis of students provided with the services. As seen in Fig. 10.13, expenditure per student at fixed prices was 8,848 TL in the public universities in 2000, while this figure decreased to 4,811 TL in 2005 (the number of students does not include open education students). Increasing to 5,145 TL in 2009, this figure was envisaged as 5,087 TL in 2010.

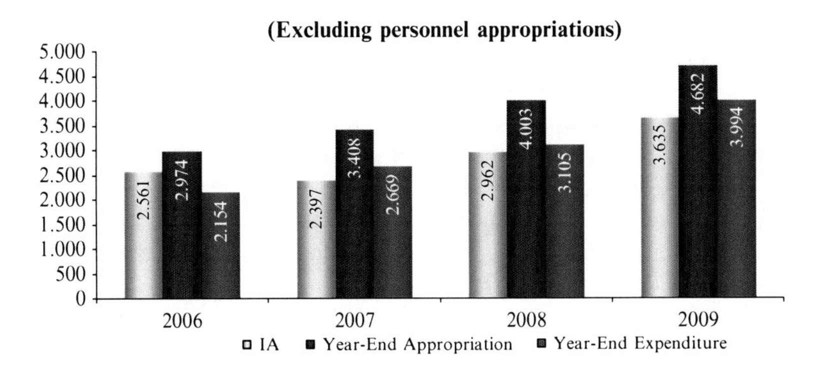

Fig. 10.14 Initial appropriations, year-end appropriations and expenditures for the public universities in 2006–2009 (million TL). *Source*: Ministry of Finance (2010)

This shows that expenditures per student have decreased considerably in real terms (Fig. 10.13). Fifty percent of the increase in the number of students (excluding open education) corresponds to the students with associate's degree (2-year undergraduate degree). It is seen that the associate's degree programs are preferred more prominently in the universities. This, in turn, causes the costs per student to shift to the areas where this impact is felt more slightly. Therefore, the system in the tertiary education tries to survive financially by following such a course. This indicator also proves that the tertiary education system focuses more on the associate's and bachelor's degrees in Turkey.

Another fundamental problem faced by the public universities in terms of financial management is failure in utilizing the resources allocated to them as their own revenues or treasury aids within the year. Although a certain part of these resources can be carried over to the following year, this does not solve the problems in utilizing the collected resources when needed. In this regard, universities fail to use 22% of the resources allocated to them, as an average of the last 4 years (Fig. 10.14). In order to find a solution to such a substantial shortfall in resource use, it will be worth to review the managerial structures of the universities.

Universities are institutions that have a separate legal entity apart from their state legal entities, as they are able to create revenues and reach significant amounts of resources. The reasons for the failure of the universities to fully use their available resources should be sought in the system itself. Considering the increase in the number of university students and requirements of today's competitive world from the tertiary education, it is quite clear that universities need more resources, since determining condition of a strong education system is undoubtedly sound financing. However, it is also necessary to manage the resources with effective, efficient, and flexible policies.

Furthermore, allocating more resources to the universities without changing the old university system would just result in waste rather than a higher-quality tertiary education service and more researches (Alesina and Giavazzi 2006, p. 69). Therefore, the question of "where has all the education gone?" should be raised in some countries where significant increases in the educational investments have no impact on the economic growth (Easterly 2001, p. 73).

Benhabib and Spiegel explored the influence of the increase in the average time of schooling of the labor force on the growth rate in their study including 124 countries for the period 1965–1985 and they did not find any correlation between the increase in per capita income and the increase in the average time of schooling. However, they concluded that human capital has an indirect influence on the economic growth through the increase in total factor productivity and this leads to an increase in domestic technology production and faster adaptation of exported technology (Benhabib and Spiegel 1994, pp. 143–173). Accordingly, the level of the human capital stock rather than the increase in the human capital plays an important role in the determination of the per capita income growth and attractiveness of the physical capital.

Pritchett suggested that the rapid growth in the human capital stock could not be reflected in the economic growth of countries such as Singapore, Korea, China, Indonesia, Angola, Mozambique, Ghana, Zambia, Madagascar, Sudan, and Senegal as a result of this study based on the data for 1960–1978. For example, although human capital stock in Zambia is very close to that of Korea, its economic growth rate is far lower than Korea. Similarly, although the enrollment rate in secondary education is 97% in the USA and 92% in Ukraine, per capita income in the USA is over 9 times higher compared to that of Ukraine (Pritchett 1996, pp. 14–44).

While in 1960, 29 countries had no university students, the number decreased to three countries (Comoros, Gambia, and Guinea-Bissau) in the 1990s. The number of students enrolled at a university increased by more than 7 times beginning from the 1960s to 1990s. The same improvement has also been seen in the other levels of education. However, this sharp increase in the number of students has not had any influence on the growth rates in many countries. Therefore, some economists argue that there is no relationship between the enrollment rate and the economic growth of a country (Easterly 2001, p. 73).

Considering these facts, it can be concluded that the contribution of tertiary education to enhancing economic development and competitiveness of the country do not depend on financial resources allocated for tertiary education. Our recommendations on increasing managerial efficiency in the tertiary education are also a result of this assumption.

Research and Development (R&D) in Turkey

R&D can assist businesses directly through the provision of skills and indirectly by way of spillovers. R&D also contributes to national development. Universities can supply the crucial underpinnings of dynamic industrial clusters. Strategies for enhancing R&D capabilities have taken a more central position in developing countries' development policy in recent years. An important step has been the promotion of university-based research and commercialization. R&D within the

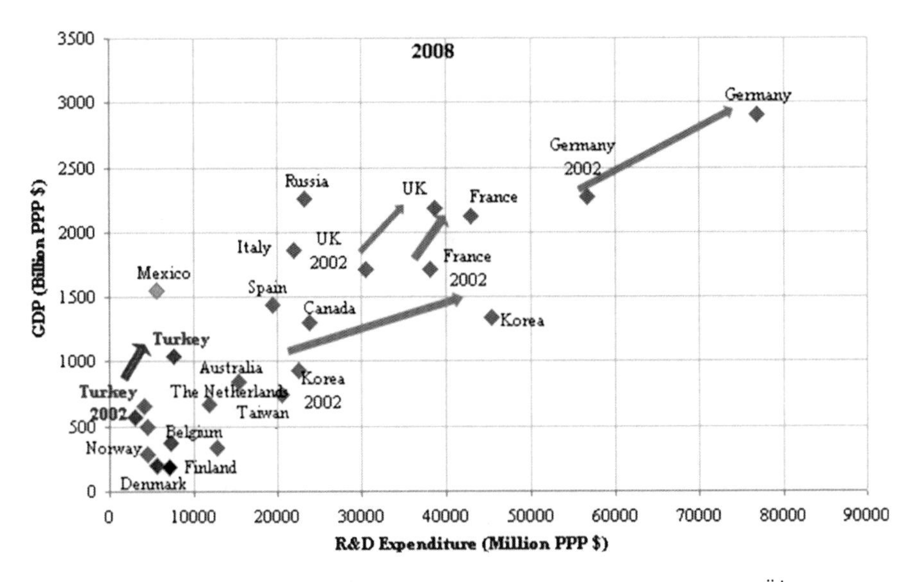

Fig. 10.15 R&D expenditure – GDP relation. *Source*: OECD MSTI, UNESCO, TÜİK

university leads important institutional underpinnings for building commercial links. Consequently, Turkey should not miss the opportunities to build bridges between the market and universities.

It is well known that R&D activities, thus the R&D expenditures have a positive effect on GDP. In Fig. 10.15, the R&D expenditures of the countries and their GDP are compared. As it is seen in the figure, there is a positive relationship between the R&D expenditures and the GDP. It is observed that the countries which have relatively high R&D expenditures have a high level of GDP. In Turkey, both the R&D expenditures and the GDP have risen as of 2002. In this respect, this positive relationship is also valid for Turkey.

There is also a positive relationship between the full-time R&D personnel and the GDP. The relationship between the full-time R&D personnel and the GDP of the countries is shown in Fig. 10.16. The GDP of the countries which have a great number of R&D personnel is relatively high, from which it is understood that there is a positive relationship here.

Turkey's average annual growth of spending on R&D was 15% between 2002 and 2006 (Fig. 10.17). On the other hand, this figure for many developed countries remained at very low rates such as it is 1.3% in France, 1.9% in the UK, 2.0% in Australia, 2.5% in Germany, 2.9% in the USA, 2.9% in Japan, 4.1% in Sweden, 6.9% in Korea, and 7.8% in Finland for the period of 1995–2005. China again had an outstanding average annual growth rate of spending on R&D with 18.5% between 2000 and 2005.

Considering the share of the R&D expenditures in the GDP of Turkey, it is clear that the rate of the resources allocated for R&D activities as a percentage of the GDP was 0.37% in 1998. This rate rose to 0.85% in terms of GDP share in 2009

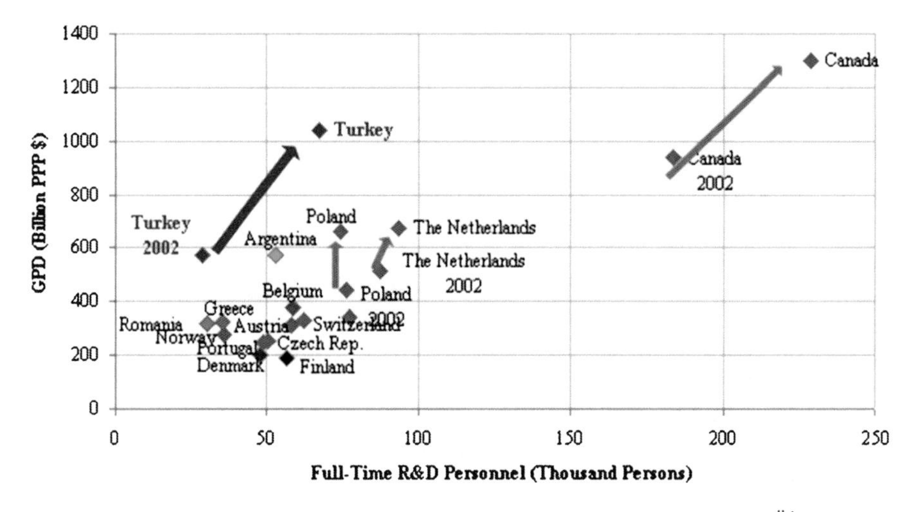

Fig. 10.16 FTE R&D personnel-GDP relation. *Source*: OECD MSTI, UNESCO, TÜİK

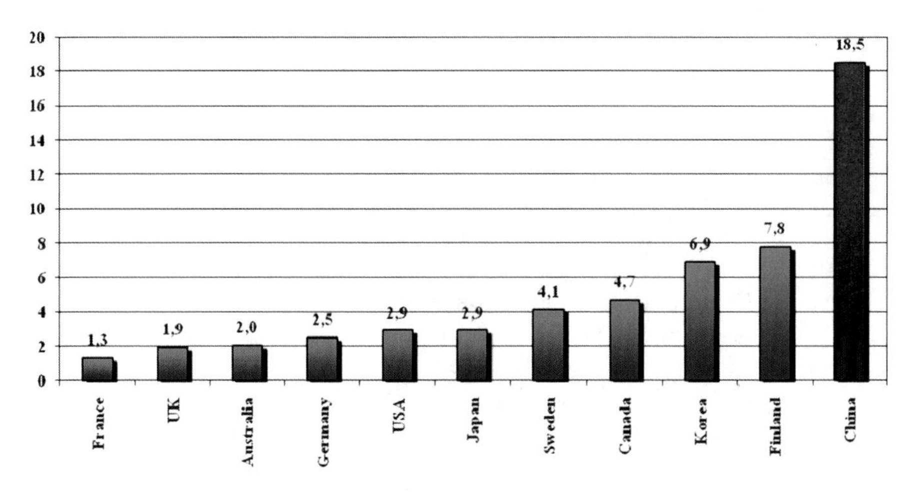

Fig. 10.17 Average annual growth of spending on R&D, 1995–2005 (%) (Constant prices, OECD. China data for 2000–2005 only) Turkey 15% (2002–2006)

(Fig. 10.18). The target is to increase this rate to approximately 2% in the following years. It is difficult to increase the share of the R&D expenditures in GDP in a short period because R&D expenditures require infrastructure. The absorption capacity of a country is important. Increasing the R&D expenditures in a country is related to a great number of factors such as tertiary education, R&D capacity of the private sector, high-quality human capital stock, and the macroeconomic infrastructure in that country. Therefore, it is expected that R&D expenditures will increase in the long run.

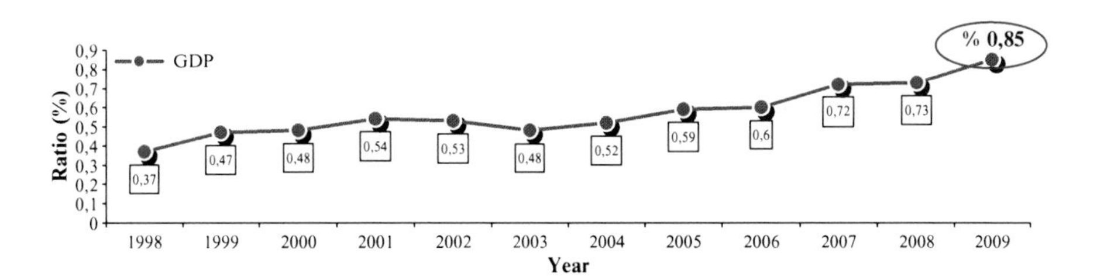

Fig. 10.18 Ratio of R&D expenditure to GDP. *Source*: TÜİK

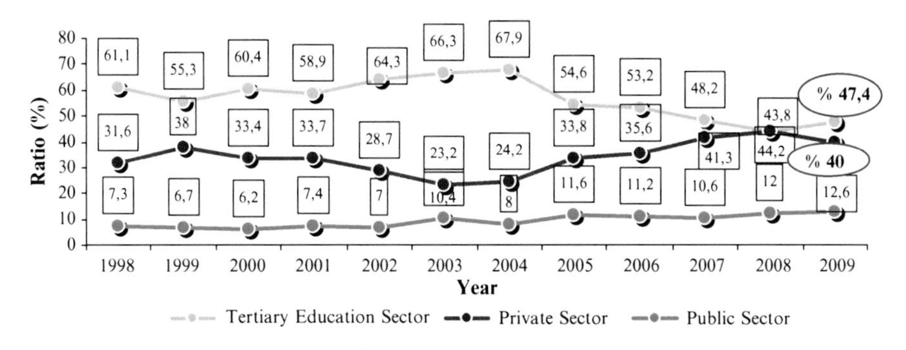

Fig. 10.19 Sectoral distribution of R&D expenditures. *Source*: TÜİK

When the distribution of the R&D investments in Turkey by the sectors is examined, it is seen that 12.6% of the total investments is made by the public sector, 40% by the private sector, and 47.4 by tertiary education institutions (Fig. 10.19).

National financial resources for the R&D activities can be listed as follows:

- TUBITAK (Turkish Scientific and Technological Research Council)
- DPT (State Planning Organization)
- Ministry of Finance
- Ministry of Industry and Trade
- University Resources
- Contract Research funded by public sector
- Contract Research funded by private sector

Furthermore, there are also some international resources available for the R&D such as the EU Framework Programs (starting from the Sixth Framework Program).

Since 2002, 26 Technology Development Zones have been created for nurturing research and innovation in the country. Currently, 19 of them are operational.

The following incentives have been offered by the Law No. 4691 under the control of Management Companies supported by the universities in which the TD Zones were established:

- Income Tax exemption
- Corporate Tax exemption
- Opportunities for Academic Staff

Thanks to the abovementioned incentives and the establishment of these Technology Development Zones, the following figures were achieved by 2008:

There are 1,154 R&D companies in these zones.

- There are 11,093 R&D personnel working in these companies.
- In total 7,280 R&D projects were launched, 4,211 of which were completed and 3,069 are ongoing.
- There are 25 foreign companies in these zones.
- USD450 million of foreign investment was attracted.
- A total of USD540 million income was generated through exports.
- Two hundred and thirty-five patents were obtained.

Outcomes

There has been an impressive improvement in the gross schooling rates, particularly in preprimary and tertiary education. In the preprimary education (3–5 ages), the gross schooling rate increased from 9.4% in the 2000–2001 academic year to 26.92% in the 2009–2010 (Fig. 10.20). In primary education, this figure exhibited an increase of approximately 5% during this period. On the other hand, the gross schooling rate in the secondary education increased from 60.97% in 2000–2001 to 84.19% in 2009–2010. Another impressive increase was experienced in the gross schooling rates of the tertiary education, recording an increase by almost 100%, from 22.25% in 2000–2001 to 44.27% in 2008–2009.

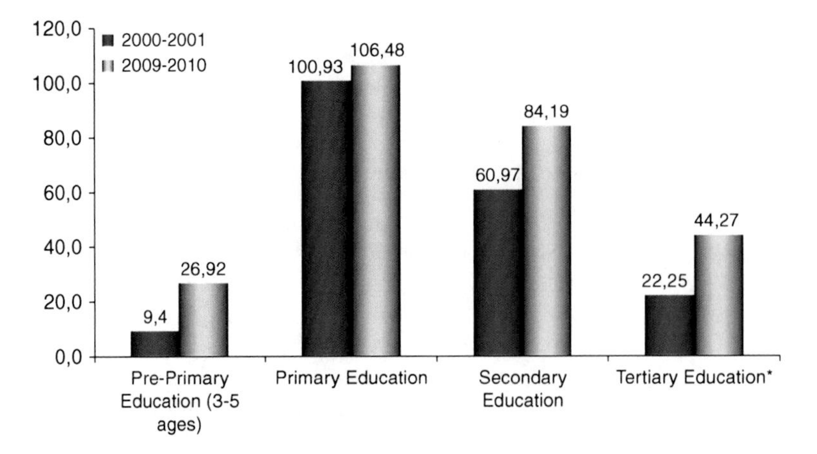

Fig. 10.20 Gross schooling rate (%).*Tertiary education schooling rates are for 2008–2009. *Source*: Ministry of National Education

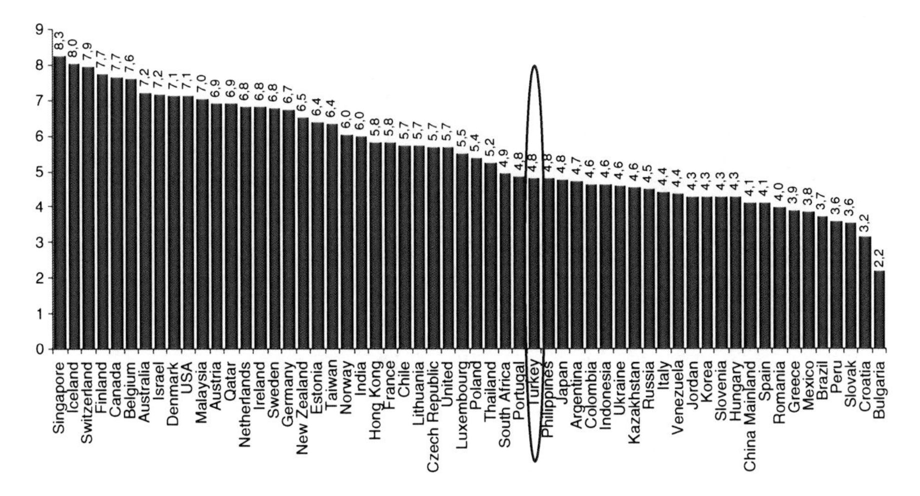

Fig. 10.21 University education meeting the needs of a competitive economy (based on an index from 0 to 10). *Source*: IMD (1995–2010)

Despite these positive developments on the schooling rates, it can be inferred from the international indexes, which compare the tertiary education system in Turkey to those in other countries, progress is still needed in the country.

It is clear that Turkey needs greater diversity in its tertiary education in order to support different regional and national needs. It also requires more efficient and selective funding mechanisms, and improved systems of governance.

In Fig. 10.21, which shows tertiary education indices, Turkey's index point is 4.8 in 2010, which is equal to that of Japan, and higher than that of Russia at with 4.5, Italy at 4.4, Korea at 4.3, and Brazil at 3.7. In 2010, Turkey ranked 34th among 58 countries in terms of this index.

In Fig. 10.22, based on an index from 0 to 10, university education attained an average of approximately 4.6 index points in 2000–2010, achieving 4.9 index points in 2000 and 2003 and 4.8 in 2010 as the highest points.

Figure 10.23 (Saygili and Cihan 2010) demonstrates the productivity levels of some countries, including Turkey, that are successful in the area of labor force productivity. It is observed here that even though Turkey was more successful in terms of labor productivity than South Korea, Chile, and Malaysia in 1975 and has made greater progress in this area since 2002 it is still behind these countries in the 32-year period. On the other hand, China and Ireland made considerable progress in labor force productivity during the same period. For this reason, as demonstrated in the figure, tertiary education should produce results which will reverse the progress in order to derive the expected benefit. In this respect, it is of utmost importance to determine the policy preferences of today and to plan a managerial structure which will put these policies into practice.

These outcomes indicate that it is time to make the necessary changes and transformations in the Turkish tertiary education system. It does not seem possible that a tertiary education system which is based on the same missions and similar

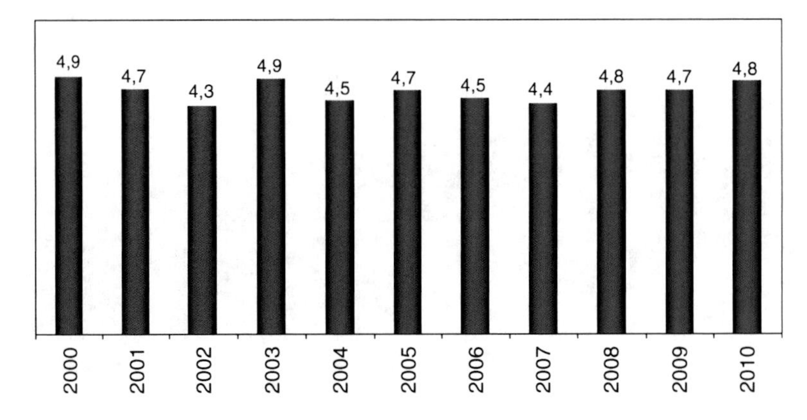

Fig. 10.22 University education meeting the needs of a competitive economy in Turkey (on an index from 0 to 10). *Source*: IMD (1995–2010)

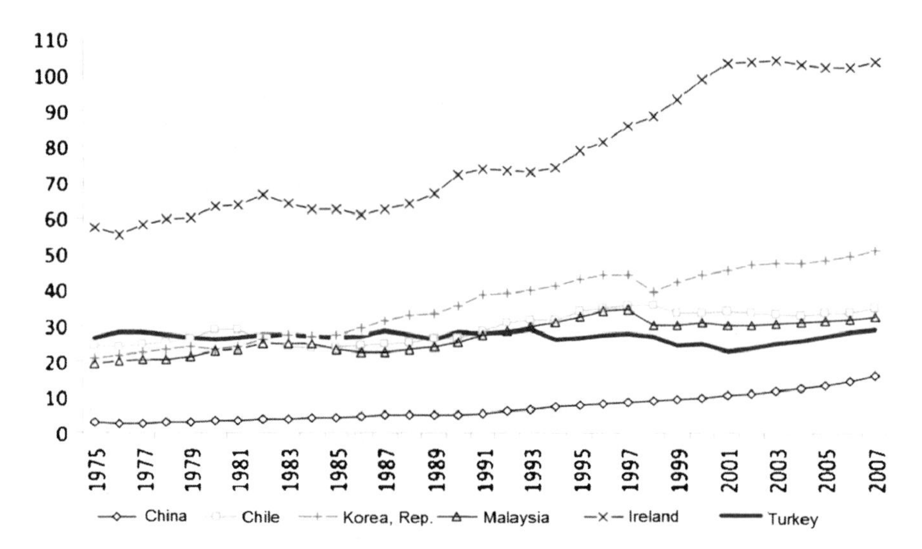

Fig. 10.23 Relative productivity level in some countries (USA = 100). *Source*: Saygili and Cihan (2010), p. 68

objectives attains the expected success in instruction, research, and service delivery to the public. Almost all universities in Turkey operate with the aim of researching, obtaining the current funds that are allocated for research, and providing graduate education. However, in terms of the requirements of this millennium, a smaller set of universities should focus their missions as research-intensive universities, and, in this context, use the funds allocated for this purpose only. A Similar case is also valid for the graduate education. For instance, more than 2,000 universities in the European Union undertake research and run graduate programs as well within the same mission. However, approximately 100 of up to 4,000 universities in the USA

Table 10.3 Performance of the national science and technology system (2006–2009)

	2006	2007	2008	2009	Target 2010	Target 2013
R&D expenditure/person (PPP $)	78	98	105	122	124	–
Number of FTE researchers (thousand)	43	50	53	58	40	–
Number of researchers/thousand	2.1	2.4	2.5	2.7	2.3	5
Public S. R&D expenditure (%)	11.2	10.6	12	12.6	12	14
Scientific publications/million people	270	311	324	348	400	400
New product turnover/total turnover (%)		9.56			10	–

Source: TÜİK

are known to be research-intensive tertiary education institutions only. As a result, the American universities figure dominantly in the league of the best research universities of the world (Lambert and Butler 2006, p. 3). (For this reason, based on their experiences, Turkey needs to support the research universities, which are few but can ably compete with their rivals.)

Turkey has made important progress because it has imposed new policies related to R&D and allocated important resources to this field. In a sense, it has started to accomplish the results of the newly imposed policies. However, Turkey has to go a long way to become a knowledge-based economy, because it is hard to accomplish the results of the investments made in R&D in the short term. Yet, in this respect, Turkey has been proceeding in the right direction.

As seen in Table 10.3, R&D expenditure per capita purchasing power parity (PPP $) which was USD78 in 2006 increased to USD122 in 2009, and is estimated to increase to USD124 in 2010. While the number of full-time researchers (engaged in full time) was 43,000 in 2006, it went up to 58,000 in 2009. The highest number of researchers (engaged in research and social work) was noted as 2,700 (in 2009), which are estimated to increase to 5,000 in 2013. Public sector R&D expenditure follows a fluctuating course. It was 11.2% in 2006; 10.6% in 2007, and 12% in 2008, and finally, it rose to 12.6% in 2009. The aim is to raise it to 14% in 2013. Scientific publications, which were 270 per million people in 2006, became 348 in 2009. The ratio of new product turnover to total turnover came as 9.56% in 2007 and it is envisaged to rise to 10% in 2010.

Table 10.4 shows that the ratio of R&D expenditure to the GDP was 0.6% in 2006, with an increase to 0.85% in 2009. The number of full-time R&D personnel was 54,000 in 2006 and it reached 74,000 in 2009. Although the private sector R&D expenditure gradually increased from 35.6% in 2006 to 44.2% in 2008, it decreased to 40% in 2009. However, it is estimated to climb up to 60% in 2013. Likewise, the private sector R&D funding recorded a gradual increase from 44.3% in 2006 to 47.3% in 2008, with a decline to 41% in 2009. Furthermore, the R&D expenditure in the tertiary education sector, which gradually decreased from 53.2% in 2006 to 43.8% in 2008, rose to 47.4% in 2009. However, this figure is estimated to decrease to 38% in 2010 and to 26% in 2013. The number of triadic patents which was 16 in 2006 is expected to increase dramatically to 100 in 2010. While the ratio of

Table 10.4 Performance of the national science and technology system (2006–2009)

	2006	2007	2008	2009	Target 2010	Target 2013
R&D expenditure/GDP	0.6	0.72	0.73	0.85	–	2
FTE R&D personnel (thousand)	54	63	67	74	–	150
Private S. R&D expenditure (%)	35.6	41.3	44.2	40	50	60
Private S. R&D funding (%)	44.3	46.2	47.3	41	50	55
T. education S. R&D expenditure (%)	53.2	48.2	43.8	47.4	38	26
Number of triadic patents	16	19	18	–	100	–
Self-innovative SME/total SME (%)	–	25.4	–	–	40	–
Co-innovative SME/total SME (%)	–	4.5	–	–	20	–

Source: TÜİK

Table 10.5 Turkey's performance in world scientific publications

Year	Ranking
1990	41
1995	34
2000	26
2005	19
2007	18
2008	17
Growth rate: 11.6% (1995–2005)	

Source: The Higher Education Council

self-innovative SME (who innovate by itself) to total SME was 25.4% in 2007, it is estimated to increase 40% in 2010. Likewise, the ratio of co-innovative SME (who innovate jointly with others) to total SME was 4.5% in 2007, whereas it is expected to increase to 20% in 2010.

Scientific publications are important tools for knowledge accumulation. In Turkey, the number of scientific publications has continuously increased since 1990. More specifically, while the number of scientific publications was 1,154 in 1990, this number increased to 6,195 in 1999 and to 22,738 in 2010. Especially after 2006, the rise in the number of scientific publications gained a considerable momentum. From a holistic point of view, the number of scientific publications increased by 1.870% between 1990 and 2008.

Turkey's ranking in the world in terms of the number of scientific publications has also increased considerably since 1990. While Turkey ranked 41st in 1990, it rose to 26th in 2000 and to 17th in 2008. In this regard, Turkey achieved an annual growth of 11.6% in terms of scientific publications in the period 1995–2005 (Table 10.5).

Turkey has a higher ranking in the world in terms of its annual growth rate for scientific publications as compared to a number of developed countries (Fig. 10.24). Turkey's annual growth rate of 11.6% is higher than that of the USA with an annual growth rate of 0.6%, Japan at 1.7%, the European Union at 1.8%, and UK with almost no increase . On the other hand, China and Korea have the two highest

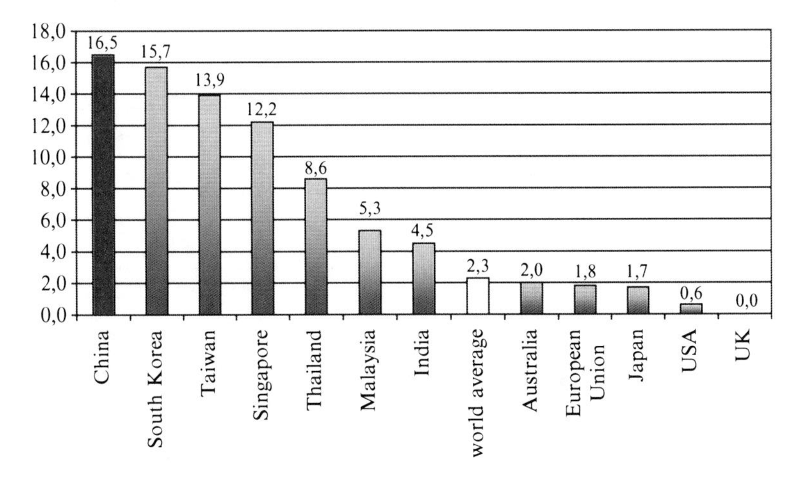

Fig. 10.24 Annual rate of growth of scientific publications, 1995–2005 Turkey 11.6%. *Source*: The Higher Education Council

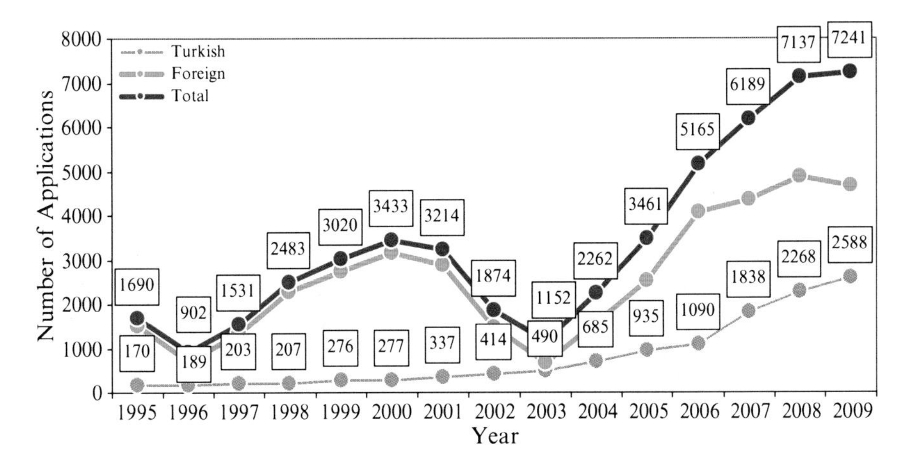

Fig. 10.25 Patent applications. *Source*: Turkish Patent Institute, R&D Statistics

annual growth rates in scientific publications with 16.5 and 15.7%, respectively. Turkey's rate is also considerably above the world average of 2.3%.

The number of total patent applications also followed a fluctuating course during the period between 1995 and 2005. It exhibited two sharp decreases in 1996 and 2003 which were triggered by the decreases in foreign patent applications. Nevertheless, it increased from 1,690 in 1995 to 5,165 in 2006 and to 7,241 in 2009. As for the domestic patent applications, the number of the applications has increased gradually and continuously from 170 in 1995 to 2,588 in 2009 (Fig. 10.25).

Although there were only 41 trademark applications in Turkey in 1995, this number increased continuously since then, reaching 3,016 in 2007. After 2007,

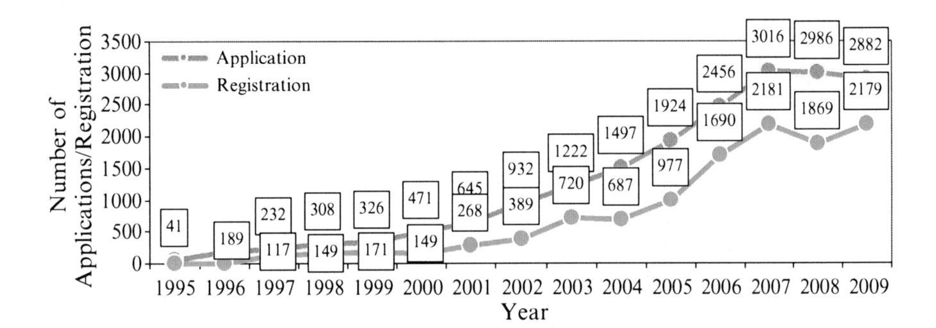

Fig. 10.26 Trademark applications and registrations. *Source*: Turkish Patent Institute, R&D Statistics

it decreased to 2,986 in 2008 and to 2,882 in 2009. All in all, the number of trademark applications rose by 6.926% between 1995 and 2009 (Fig. 10.26).

However, there seems to be no registration in Turkey in 1995 and 1996. In 1997, the number of registrations increased to 117 and to 720 in 2003. After 2003, Turkey reached the highest level by 2,181 registrations in 2007 while this figure decreased to 1,869 in 2008 and rose back to 2,179 in 2009.

Conclusion

In recent years, the tertiary education system in Turkey has improved significantly in terms of numbers and resources. Important innovations regarding instruction and R&D have been put into effect. These are major developments for the tertiary education system. However, in the process of making improvements, new challenges have arisen for the Turkish tertiary education system. New policies in the tertiary system should focus on enhancing the quality of education, which can occur in a number of ways.

First, the universities should make the arrangements concerning their managerial structures in accordance with the generally accepted standards, thereby allowing the universities to establish their own effective management models. Good governance will also advance both freedom and accountability, which is a prerequisite for increased funding from both the public and the private sectors.

In addition, the universities should be able to transfer technology from the campus to the market. In this respect, a regime for intellectual property right protection should be established and shaped in such a way that resource transfer enables knowledge transfer. Thus, in the area of research resources, priority should be given to the universities and researches, which are the most competent in this field and can ensure technology transfer.

Turkey also needs to focus on being more competitive on an international level. International competitiveness depends on the success level of the quality of education and scientific researches. In this sense, achievement of medium and long-term economic growth is correlated with the effective functioning of the tertiary education system.

Furthermore, the universities need to understand how to produce public value and service with the obtained resources, by following which processes and management policies create this value and to internalize them at institution level. The universities should also be able to operate a process through which they will attain their goals and objectives, carry out activities, and provide services undertaken with these activities. Otherwise, they cannot assume the public responsibility required in the areas such as instruction, scientific research, and service delivery to the public at desired level.

Thus, improved governance and increased financial and managerial autonomy are both critical issues for Turkish universities. Unless they can demonstrate that they can run their affairs efficiently, universities will not be able to build a credible case for more funding. Additionally, they must have the authority to decide their own strategy for teaching and research, if they hope to flourish in the increasingly competitive environment.

Finally, universities are the institutions, which have the capacity and responsibility to implement the processes and the mechanisms for the effective and efficient functioning of managerial and financial affairs. Consequently, there is no doubt that universities need more resources. Therefore, sound financing and managing resources with good policies are preconditions for a quality educational system.

References

Alesina A, Giavazzi F (2006) The future of Europe. MIT Press, London

Benhabib J, Spiegel MM (1994) The role of human capital in economic development: evidence from aggregate cross-country data. J Monetary Econ 34(2):143–173

Easterly W (2001) The elusive quest for growth: economists' adventures and misadventures in the tropics. MIT Press, Cambridge

Emil F, Yilmaz HH (2003) Kamu Borçlanması İstikrar Programları ve Uygulanan Maliye Politikasının Kalitesi (Public borrowing, stabilization programs and the quality of the government financing policy being applied). METU ERC Working Paper 03/07

International Institute for Management Development (2010) World Competitiveness Yearbook, 1995–2010

Korkmaz E (1975) Yüksek Öğrenimde Etkinlik (Efficiency in high education). Basılmamış Doktora Tezi (unpublished Ph.D. Thesis), Istanbul

Lambert R, Butler N (2006) The future of European universities: renaissance or decay? Centre for European Reform, London

Maliye Bakanlığı, Muhasebat Genel Müdürlüğü, Kamu Hesapları Bülteni Sayfası (web page of Ministry of Finance, Directorate General of Accounts, Bulletin of Public Accounts). Erişim Mayıs 2010, www.muhasebat.gov.tr. Accessed May 2010

Organisation for Economic Co-operation and Development (2009) Education at a glance. OECD, Paris

OSYM İstatistikleri (Statistics of Student Selection and Placement Centre). Accessed 11 May 2010

Pritchett L (1996) Where has all the education gone? The World Bank Policy Research Working Paper 1581
Salmi J (2009) The challenge of establishing world-class universities. The World Bank, Washington
Saygili Ş, Cihan C (2010) Dünyada ve Türkiye'de Verimlilik: Karşılaştırmalı bir Analiz ve Türkiye'nin Perspektifi (Productivity in the world and in Turkey: a comparative analysis and the perspective for Turkey). İşveren Dergisi 68:67–70
Turkish Statistics Office (2002) Household spending on education survey results (TÜİK 2002 Hanehalkı Eğitim Bütçe Anketi Sonuçları)
WBI (2007) Building knowledge economies. WBI, Washington, DC
Yilmaz HH (2007) İstikrar Programlarında Mali Uyumda Kalite Sorunu: 2000 Sonrası Dönem Türkiye Deneyimi (The problem of quality in the stabilization policies and financial consistency: Turkish experience after 2000). TEPAV, Ankara
Yilmaz HH, Kesik A (2010) Yüksek Öğretimde Yönetsel Yapı ve Mali Konular: Türkiye'de Yüksek Öğretimde Yönetsel Etkinliği Artırmaya Yönelik Bir Model Önerisi (Administrative structure and financial issues: a proposed model to enhance administrative efficiency in higher education in Turkey). Maliye Dergisi 158:124–163
Yilmaz HH, Emil F (2008) Social expenditures at different levels of government: Turkey. A background study prepared for World Bank Social Policy Work

Websites

http://nsse.iub.edu/html/about.cfm. Accessed May 2010
http://www.admin.cam.ac.uk/univ/annualreport/. Accessed May 2010
http://www.admin.ox.ac.uk/rso/statistics/ar_2008-09.shtml. Accessed May 2010
http://budget.asu.edu/state-investment-0. Accessed May 2010
http://introduction.ku.dk. Accessed May 2010
http://www.calstate.edu/business_community/. Accessed May 2010
http://www.nih.gov/about/NIHoverview.html. Accessed May 2010
http://www.unistra.fr/uploads/media/recettes_uds.pdf. Accessed May 2010
http://www.urmc.rochester.edu/news/publications/annual-report/2009/research/nih-funding.cfm. Accessed May 2010

Part III
Industry Case Studies

Chapter 11
Local Content Rules as a Tool of Technology Transfer in the Turkish Rolling Stock Manufacturing Industry: Tulomsas Experience

Muammer Kantarci

Introduction

In Turkey, passengers and goods are transported mainly by road reflect a very low share of railways in total freight (5%) and passenger transport (3%). A main reason for this fact is low railway density (length of railways per square kilometer of territory) in the country. It is clear that over time the rise in transport demand due to the increase in population and economic growth cannot be possibly be met by the existing road network. Therefore, the share of railways in total transport market is likely to increase. Changes in government policy since 2003, as reflected by the "Transportation Master Plan Strategy" (2005) and the "Transportation Infrastructure Need Assessment Study (TINA)," will assist that process.

Under the new policy, rehabilitation of existing tracks and rolling stock as well as new High Speed Rail (HSR) projects, new generation projects on locomotives and modernisation projects like signaling and electrification on the existing routes have been initiated. That would translate into increased demand for locomotives and other rail vehicles. The development local technological and manufacturing capacity is thus crucial both to benefit from increasing demand for rolling stock as well as assisting the government policy in increasing the railway sector's role.

This chapter presents a case study of a successful technology transfer agreement between a locomotive manufacturer from the USA (EMD) and a local Turkish locomotive manufacturer, TULOMSAS (Fig. 11.1). The licensing agreement converted a standard procurement process into one where the local content rule helped TULOMSAS to manufacture about half of the total contract price locally. Additionally, an offset provision helped the local manufacturer to export products to South Africa as part of an EMD contract.

M. Kantarci (✉)
Turkish Coach Industry Incorporation (TÜVASAŞ), affiliated company of TCDD, Adapazari, Turkey
e-mail: dr.kantarci@gmail.com

M.A. Yülek and T.K. Taylor (eds.), *Designing Public Procurement Policy in Developing Countries*, DOI 10.1007/978-1-4614-1442-1_11,

Fig. 11.1 Final DE 33000 mainline locomotive

The chapter starts out by giving general definitions of technology transfer and introducing the licensor and the licensee. It than presents the details of the technology transfer transaction.

Prime Methods of Technology Transfer

Technology can be defined as the body of knowledge, tools, techniques, and innovations derived from science and practical experience that is used in the development, design, manufacturing and application of products, processes, systems, and services.

Technology Transfer is the process of sharing skills, knowledge, technologies, methods of manufacturing, samples of manufacturing, and facilities among governments and other institutions to ensure that scientific and technological developments are accessible to a wider range of users. The latter can then further develop and exploit the technology into new products, processes, applications, materials, or services

Technology transfer methods can be classified into direct and indirect categories. Direct Technology Transfer can be realized through:

- Foreign Direct Investment
- Joint Ventures
- Technology Transfer Agreements
  - License Agreements
  - Supply Management Agreements
  - Technical Support Agreements

  - Turn-key Contracts
  - Distributor License Agreements

- Machinery and Equipment
- Financial Leasing
- Employment of Foreign Experts
- Free Zones
- Subcontracting
- R&D Activities

Indirect Technology Transfer can on the other hand, primarily be realized through:

- Public Knowledge
- Education and trainings
- Human Resources
- International movement of people

Licensing Agreements for Technology Transfer: Some Guidelines

Licensing Agreements (LA) are among the direct technology transfer methods. In drawing up a successful License Agreement, best practices suggest the importance of the following success factors[1]:

1. Preliminary steps
   - Compiling project information
   - Selection of technology
   - Selection of licensor
2. Provisions of the Technology License Agreement
   - Description of technology and technical assistance
   - Technologies and their terms of transfer to be clearly covered in the Agreement
   - Access to improvements during the Agreement period
   - Guarantees
   - Remuneration: The total amount of payments over the period of Agreement should be carefully assessed, such as technology providers interests to protect their intellectual property versus developing country partners' interests in eliminating royalties and fees
   - Duration issues

[1] UNIDO (1976).

- Sales territory, exportation and exclusivity issues: the licensee should seek to obtain an exclusive right for manufacture and sale at least within the receiving country
- Patent and trademark issues
- Sublicensing rights
- Confidentiality
- Procurement of components and intermediate products
- Currency provisions
- Assignability
- Training: The agreement should provide for adequate training in the licensor's work and facilities and in-plant training in the licensee's plant. In the case of the former, the number of persons to be trained, the areas of the training and their duration, together with arrangements to be made for the training, should be clearly defined in the Agreement
- Expatriates and local managers should be in a good mix in order to facilitate technology transfer and cultural exchange and change
- The most-favored-licensee clause
- Termination
- Governing law and language
- Arbitration
- Force majeure clauses

Licensee Overview: TULOMSAS

Company Overview

TULOMSAS is a joint-stock company formed under the Turkish Commercial Code. It is subject to state financial auditing. The highest governance body is the Board of Directors consisting of the Chairman and four members. The General Manager of TULOMSAS is at the same time the chairman of the Board of Directors.

As an affiliated company of the Turkish State Railways, TULOMSAS met all of Turkey's locomotive and wagon needs until 1985, when it took on a new identity as the Turkish Locomotive and Engine Industry, Inc. Subsequently, it developed into a major modern manufacturer with customers across the Middle East and the Balkans, accumulating considerable experience in building locomotives, freight cars, diesel and electric engines for locomotives, light-rail rolling stock, tank engines, traction engines, and other types of heavy industry products. The company serves the needs of TCDD (Turkish Railways Administration) as well as exporting to other countries.

TULOMSAS production is realized in six different factories (Locomotive Factory, Wagon Factory, Diesel Engine Factory, Electrical Machinery Factory, Gear & Tooling Factory, Foundry & Chemicals Process Factory). Since "Karakurt" the first in-house designed locomotive in 1961, TULOMSAS has manufactured various other in-house designed locomotives (such as DH 7000, DH 9500, and DH 10000 shunting locomotives).

TULOMSAS is also a leading Maintenance Repair and Operations (MRO) services provider in Turkey and its region.

Company Milestones

During the construction of the Baghdad railway in the late-nineteenth century, a small maintenance and repair workshop named the Anatolia-Ottoman Rail Company was established in Eskişehir in 1894 by the German government to realize the repair and maintenance of steam locomotives and wagons used in the Anatolia-Baghdad Railway Line. This constituted the foundation of today's TULOMSAS. Small-scale repairs of locomotives, coaches and freight cars have been done in this workshop, while the boilers of steam locomotives were sent to Germany for repair. All spare parts were imported from other countries.

In 1923, the small repair workshop had a closed area of 800 m. Many new units for producing steam locomotive boilers, gears, a carpenter's shop, other units producing materials for railway bridges, switches, and tracks were established between the years 1925 and 1928 with aim of reducing the dependency on other countries. The annual maintenance and repair capacity of the railway repair shop increased in those years to 3–4 locomotives and 30 passenger and freight cars.

In 1961 the first Turkish steam locomotive "KARAKURT" with 1,915 HP, 97 t in weight and 70 km/h speed was manufactured.

In 1971 the first diesel–electric mainline locomotive with 2,400 HP, 111 t weight and 39,400 kg traction force was manufactured under the license agreements with French Traction Export Company for locomotive and with Chantiers de L'Atlantique and were put into service. DE 24000 type Diesel Electric Locomotives have been produced up until 1985.

In 1986, the first DE 11000 type mainline and shunting locomotive with 1,100 HP was manufactured under the license agreement with German Krauss-Maffei Company for the locomotive and with MTU Company for the diesel engine. 70 units of the DE11000 type locomotives were produced up until 1990.

In 1987, TULOMSAS started to build the first DE 22000 type mainline locomotive with 2,200 HP under the license of the EMD General Motors Company/USA. Thirty-nine units of the DE 22000 type mainline locomotives were imported by the TCDD and 48 units were manufactured in TULOMSAS.

Moreover, the production of various railway work machines (snow-plows, railway mobile cranes, track maintenance cars with light crane, catenary maintenance cars) started in 1987. In 1988, TULOMSAS initiated the production of the E 43000 type electric mainline locomotive with 4,300 HP under the license agreement with the Japanese NISSHO IWAI-TOSHIBA company. Forty-four units were manufactured in TULOMSAS after importing one complete locomotive from Japan.

Table 11.1 Technology Transfer Experience of TULOMSAS

Partner	Product	Method	Initial agreement year
General motors, EMD/USA	DE 22000 and DE 33000 Mainline locomotive	Technical assistance	1984 and 2001
TOSHIBA Corp./ Japan	E 43000 electrical mainline locomotive	License	1985
ALSTOM/France	Diesel engine	License	1980
MTE/FRANSA	Locomotive production	License	1980
Krauss-maffei/ Germany	Locomotive mechanical equipment	License	1986
MTU/Germany	Diesel engine	License	

In 2003, TULOMSAS made an agreement for technology transfer with the Electro-Motive Division of the General Motors/USA (which now is active as the Electric Motive Diesels Company) and manufactured the first six units of DE 33000 type diesel–electric mainline locomotives to meet part of the need of Turkish State Railways for 89 mainline locomotives. Twelve units of the remaining 83 locomotives were built in 2004. The remaining 71 locomotives will be manufactured with 51% local content.

Consequently, TULOMSAS gained experience for over a century by building locomotives within the scope of the existing license agreements and using the technology of international trade marks like GM, Toshiba, Krauss-Maffei, MTU, and Semt Pielstick.

Technology Transfer Experience of TULOMSAS

TULOMSAS has been manufacturing various types of locomotives under licensing or technical assistance agreements with reputable international companies (Table 11.1).

Licensing agreements or technical assistance has led to increasing local content. Manufacturing of DE33000 type diesel–electric mainline locomotives built in cooperation with the EMD Company (USA) are a case in point (Fig. 11.1). As a result of the relevant technology transfer process, local content in the manufacturing of the DE 33000 locomotives has increased from 20% in the first years of operations to 51% currently. That corresponds to savings and value added for the local economy.

Licensor Overview: EMD

Harold L. Hamilton and Paul Turner founded the Electro-Motive Engineering Company in Cleveland, Ohio, in 1922. In 1925, the company changed its name to Electro-Motive Company (EMC) and entered full-scale production, selling 27 railcars.

In 1930, General Motors, seeing the opportunity to develop the diesel engine, acquired EMC. In 1941, General Motors merged the EMC and Winton Engine Company (another GM subsidiary) to form GM's Electro-Motive Division (EMD-GM).

By the end of the war, under growing demand, diesel driven locomotive production of EMD-GM increased significantly with the new passenger EMD E-units and the new improved freight locomotive the EMD F3 following in late 1946. That led to the opening of a new locomotive production facility at Cleveland, Ohio in 1948 to meet the demand for diesel locomotives.

In 1949, EMD-GM opened a new plant in Ontario, Canada, which was operated by the subsidiary General Motors Diesel (GMD), producing the existing EMD, as well as unique GMD designs for the Canadian domestic and export markets.

In 2004, GM had agreed to sell the EMD-GM to a partnership led by Greenbriar Equity Group LLC and Berkshire Partners LLC which changed the name of the company to "Elecro-Motive Diesel (EMD)". In 2010, the Progress Rail Services Corporation, a subsidiary of Caterpillar, Inc. acquired EMD from the previous owners.

The EMD currently maintains major facilities in LaGrange, Illinois, United States, in London, Ontario, Canada, and in San Luis Potosí, Mexico. The EMD plans to open an additional facility in Munice, Indiana, and United States in late 2011.

The EMD holds approximately 30% of the market for diesel–electric locomotives in North America, second only to its competitor the GE Transportation Systems, which holds the remaining 70% share of the North American market.

The Case Study of a Licensing Agreement: DE 33000 Mainline Locomotives Technology Transfer Agreement Between EMD and TULOMSAS

EMD[2] and TCDD and TULOMSAS have been in cooperation since the early 1980s. The joint locomotive production between EMD and TULOMSAS has now a history of over two decades and continues in connection with the manufacturing

[2] EMD has changed ownership over time. The initial partner of TCDD was EMD-GM. Following the ownership change in 2004, the name of the company became EMD. We use "EMD" and "EMD-GM" interchangably in this chapter.

of new generation locomotives, major locomotive components and the locomotive underframes.

The cooperation involved supply of locomotives by EMD to TCDD augmented by technology transfer from EMD to TULOMSAS through licensing agreements with conditions of minimum local content. The cooperation between the two parties has developed in two phases each corresponding to a separate procurement. In the second stage, the minimum level of local content has been increased considerably. In addition, an offset provision, involving parts to be exported by TULOMSAS to South Africa, was introduced in the second stage.

Ultimately, the technology transfer assisted TULOMSAS to develop local technological and manufacturing capacity and becoming a supplier of EMD in the latter's sales to third countries. Thus, local content rule has been instrumental in increasing the capacity of the local manufacturer and assisted it in increasing its exports.

Technology Transfer from EMD to TULOMSAS: The First Phase

In early 1984, TCDD signed an agreement with EMD-GM, the predecessor of EMD to procure thirty-nine units of DE 22000 type mainline locomotives. The provisions of the agreement also covered licensing, under which, TULOMSAS was to manufacture forty-eight DE 22000 locomotives. TULOMSAS built the first DE 22000 locomotive in 1987, subsequently increasing the manufactured units to forty-eight. The local content was agreed at twenty percent, meaning the share of local content in the unit contract price of the forty-eight locomotives was to be twenty percent.

The agreement signed between the Turkish State Railways-TCDD and The Electro-Motive Division of General Motors Corporation included:

- Technology Transfer Agreement
- License Agreement
- The Contract for purchasing of diesel–electric main line locomotives
- The Supply Agreement
- Technical Assistance Agreement
- Bank-funded loan facility

Technology Transfer from EMD to TULOMSAS: The Second Phase

In 2001, TCDD and EMD-GM (which subsequently became the "Electric Motive Diesel" company in 2004) signed a new agreement for the procurement of seventy-one DE 33000 mainline locomotives. In 2003, when a new management took over at TULOMSAS, a new round of negotiations were launched and rapidly finalized.

Table 11.2 Local content and cost distribution

	Number of locomotives	Unit cost (USD)	Foreign content	Local content	Total cost (USD)
First package (financed by local funds)	18	2,215,121	76.1%	23.9%	39,872,178
EMD-GM	18	1,685,000	76.1%		30,330,000
TULOMSAS	18	530,121		23.9%	9,542,178
First package (financed by bank credit)	24	2,341,704	50.2%	49.8%	56,200,896
EMD-GM	24	1,175,661	50.2%		28,215,864
TULOMSAS	24	1,166,043		49.8%	27,985,032
Second package (financed by bank credit)	47	2,339,995	45.7%	54.3%	109,979,765
EMD-GM	47	1,069,995	45.7%		50,289,765
TULOMSAS	47	1,270,000		54.3%	59,690,000
Total package	89	2,315,200			206,052,839

On July 17, 2003 Area Manager/International Locomotive Sales sent to Dr. Muammer Kantarci, TULOMSAS CEO, a letter pertaining to the discussions concerning cooperation between the two parties. The letter focussed on three main areas: Contract Maintenance, Offsets, and Locomotive Exports to the region. Selected excerpts from the content of the letter, presented in Annex-I show the main lines of the licensor's commitments required by the potential licensee. This letter led to a Protocol signed on November 10, 2003.

The Protocol (Annex-II) set out the details of the local content rules governing the cooperation between TULOMSAS and GM-EMD. The Protocol provided that fifty-three of the total of seventy-one locomotives were to be manufactured with 51% local content. Subsequently total number of locomotives was raised to eighty-nine. Moreover, the agreement included an offset provision whereby TULOMSAS was to supply fifty units of GT 26-MC locomotive underframes as part of EMD locomotive exports to South Africa.

Under the Protocol, all of the eighty-nine DE 33000 locomotives were ultimately manufactured by TULOMSAS under two packages (Table 11.2). The build-up started in 2004 and was completed by 2009. Average local content increased during this period reaching 47.2% in the total procurement (eighty-nine units) and 53.8% in the original number of units (seventy-one units).

Conclusions

Technological development is crucial for sustained growth performance of the Turkish economy. Railways are one of the major sectors with a significant growth potential and where technological development can positively spillover to other sectors.

Licensing agreements constitute an important tool for rapid and sustainable technology transfer in complex and technology-intensive fields. Licensing, if designed and implemented effectively, can be instrumental in transferring the body of knowledge consisting of know-how, best practices and general experience. It can be efficiently used to close the gap between high and low technology entities as well as assisting the licensee in gliding down the learning-curve.

Strategically, Turkey aims at becoming a logistical hub between Europe and the Balkans, Middle East, Russia, C.I.S, Caucasia, Black Sea, and Mediterranean countries. Railway projects are expected to play a significant role in the realization of this strategic goal as included in the "Transportation Master Plan Strategy" (2005) and the "Transportation Infrastructure Need Assessment Study (TINA)."

A more pronounced role for railways makes low-cost manufacturing and MRO services of rolling stock a prime area in need of technological development. TULOMSAS experience with technology transfer has so far been successful. This chapter has summarized the main tenets of the technology transfer process between EMD and TULOMSAS. This experience indicates that successful design and implementation of a licensing agreement has contributed to the development of local manufacturing capacity, currency savings, and export earnings.

Annex-I: Excerpts of the Letter from Area Manager/Locomotive Sales, GM-EMD addressed to CEO of TULOMSAS Dated July 17, 2003

Contract Maintenance

I know from our meeting, facility upgrade is one area of primary interest. As with all aspects of our programs, the facility upgrades required and provided, have varied widely depending on the requirements. Anywhere from mere tooling refreshes to extensive facility overhaul or even construction of new facilities. The actual amount for each program depends on detailed facilities assessments and investigations for which EMD has specialists on staff.

The most common areas of improvement are:

- Lighting
- Drainage systems/oil recovery
- Compressed air, water, lube oil access
- Overhead cranes
- Wheel true machines
- Drop table or jack stands
- Warehousing
- Locomotive wash stations

As a preliminary step prior to an actual visit to Eskisehir, TULOMSAS could assist EMD-GM by providing a detailed analysis of TULOMSAS' facilities and any other general requirements. Some of the areas/items we are interested in are as follows:

- Locomotive Duty cycle?
- Hours of operation – 5 day operation or 7 day 24 h operation?
- Miles or kilometers per month?
- Typical reliability measurements used?
- Typical availability measurement used (calculation)?
- Total number and location of shops to be used for maintenance?
- Equipment at each shop (cranes, drop tables, special tooling)?
- Load test facilities?
- Inspection pits?
- Is the workforce unionized?
- What is the skill level of the assigned workforce?
- Describe the warehouse facilities as to size, security, availability?

To further facilitate this we have attached a "Questionnaire which TULOMSAS can complete and return to EMD/GM."

Offsets

"I have not completed my full review into this area. I do not have the specifics at this time but as part of my discussion with our material procurement coordinator, I discussed the history of our attempts in the past to have TULOMSAS become our supplier. In addition, I will continue my review, and I am sure there will be other possibilities which we can include for future discussions."

Locomotive Exports

This is a very interesting and valuable area of cooperation for all parties. In addition to countries like as, neighbor countries where we have already cooperated, there are many more in the Middle East, Africa, and Asia which will be in need of reliable and economical locomotives in the future. Probably one of the biggest potentials is in [IRAK]. TULOMSAS' demonstrated capabilities with, and the performance of, both the GT26CW-2 locomotives now being manufactured for the TCDD, along with the diesel hydraulic shunters you have designed, manufactured, and supplied for the IRR, will influence their future decision(s).

There is no doubt there is a large potential market here. With the proper mutual cooperation there should be a way to get TULOMSAS manufactured EMD locomotives into these markets.

Annex-II: Protocol Between TULOMSAS and GM-EMD dated May 8, 2002 Specifying the Local Content Rule

With reference to the Contract no.03/0.56-1 dated 08.05.2002 signed between TCDD General Directorate (TCDD) and General Motors Corporation-Electro Motive Division (GM-EMD) for the procurement of 53 each Model GT26CW-2 Diesel Electric Mainline Locomotive CKD kits for local manufacture by TÛLOMSAŞ General Directorate (TULOMSAŞ) under the existing license agrément, in order to increase the scope of the local content defined in Enclosure-1 and already achieved by TULOMSAS in the manufacture of 6 each locomotives of the same model for TCDD, it has been agreed by both parties that the local content of TÛLOMSAŞ will be increased to 51 % by the localization of the additional main components, works and processes stated in Enclosure-2.

Within the scope of this protocol, all the manufacturing drawings and technical specifications for the said main components, works and processes have been delivered by GM-EMD to TULOMSAS free of charge under the existing license agrément and technical support and training will be similarly provided free of charge during the manufacture of the locomotives under the terms of the above contract.

The licensor GM-EMD has furthermore stated that due to the current expiry date (31.12.2003) of the Canadian Government's Export Agency's (EDC) credit offer submitted on 08 May 2002 for the financing of this project, the affectivity of this protocol will be conditional to the initiation of the loan agrément negotiations between the creditors and the Turkish Republic's Undersecretariat of Treasury prior to 31.12.2003.

Enclosure 1

1. Underframe Manufacture and Machining.
2. Car body and Driver's Cab Manufacture.
3. Equipping, Piping and Cabling of the Underframe, Cab and Carbody.
4. Fabrication of Main Air Reservoir.
5. Fabrication of Battery Box.
6. Fuel Tank Fabrication.
7. Assembly of Control Desk.
8. Assembly and Testing of Brake System.
9. Fabrication and Assembly of Engine Equipment Rack.
10. Manufacture of Buffers, Drawgear and Automatic Couplings.
11. Final Assembly of Locomotive.
12. Painting of the Locomotive.
13. Final Testing of the Locomotive.

Enclosure 2

1. Fabrication, Equipping, Wiring and Production Test of the High Voltage Cabinet.
2. Manufacture of Bogie Components like Axles, Wheels, Rubber Elements, Bearing Adaptors and Brake Rigging.
3. Casting of the Bogie.
4. Machining of the Bogie Frame Casting.
5. Assembly and Test of the Bogie Subassemblies and Bogie.
6. Fabrication of Equipment Rack Components like Auxiliary Control Cabinet (AC Cabinet).
7. Equipping and Testing of the AC Cabinet.
8. Fabrication of the Water Tank.
9. Fabrication or Manufacture of the Traction Motor Blower Housing, Dust Bin Blower Housing, Auxiliary Generator Mounting Brackets, Compressor Drive Shaft and Auxiliary Generator Drive Shaft.
10. Assembly and/or Partial Manufacture and Testing of the Traction Motors.
11. Traction Motor and Generator Power Cables.
12. Driver's Cab Interior and Exterior Design Modification with Engineering Support from GM-EMD and Fabrication and Equipping and Cabling of the Driver's Cab.

The protocol has been approved by the TCDD Board Directors in August 2004.

Definitions: Technology Transfer

"The Technical Information '' shall mean drawings, specifications and other documentary technical information furnished to TULOMSAS-TCDD by EMD-GM under License Agreement.

Main Diesel Locomotives' shall mean vehicles, designed for operation on rails are powered by GM Model 645 Series diesel Engines with and AR10/D18 main Generators and current model D78B traction motors designed by EMD primarily for use in Diesel Locomotives and intended primarily for propulsion."

Product in Technology Transfer

Main Characteristic of Diesel Electric Locomotive that have been subjected to Technology Transfer Agreement in Table 11.3 below:

Table 11.3 Main characteristics of the product

Weight of locomotive	119 t
Axle load	19.83 t
Axle arrangement	Co–Co
Speed	131 km/h
Gage	1,435 mm
Wheel diameter	1,016 mm (new)
Drive system	Diesel Engine + Alternator + Redresser + Traction Motor
Fuel tank	6,435 L (1,700 Galon)
Diesel engine type	16-645E3C (Turbocharged)
Engine power and speed	2,463 kW/3,300 HP – 904 rpm
Idle speed	318 rpm
Number of cylinder	16
Cylinder diameter	230.19 mm (91/16″)
Cylinder stroke	254 mm (10)
Cylinder arrangement	V45°
Type of traction motor	D78B
Main generator	AR10
Auxiliary generator	CA 6A
Minimum curve radius	200 m

The Issues for Implementing the Technology Transfer

Although it was mentioned earlier that TULOMSAS and EMD-GM joint locomotive production has a history going back to 1987, TULOMSAS has to improve and develop its capability of manufacturing the new product.

The Electro-Motive has an ongoing development program focused on the continuous development and implementation of new manufacturing strategies and processes. This program utilizes the latest technologies, process development and statistical analysis to optimize locomotive maintenance. Their overriding philosophy is to use a combination of techniques, including condition-based maintenance, on-board testing, remote monitoring, statistical analysis, process development and others to meet the objectives.

In order to focus on continuous improvement, there are a few underlying goals that drive the development activities. Both parties have agreed to follow up the following issues;

- Quality Issues
- Technical Issues
- Localization Issues-Bogies
- Commercial Issues-Offset

Quality Issues

- Supplier Quality Questionnaire Assessment.
- Who should fill the application (EMD or TULOMSAS)?
- Product/Process Warranty.
- First article Inspection – full dimensional and welding check for the first two under frames.
- The first two under frame units shall be built under supervision of EMD and will be checked for 100%.
- Process Flow Diagram – completed has casting and underframe production flow chart (TULOMSAS will send casting flow diagram).
- Inspection and Test Plan – Control and Welding Plan.
- Functional/Performance – This one will cover welding process – training, WPS, WI.
- Material Test Results – The approval of the substitutions for material requirements.
- When completed, the list of all the approved substitutions should be compiled (with EMD approvals) and attached to SPAP documentation.
- Will supply vendor certificates.
- Will 100% test for the two under frame, remainder will provide samples results.
- Process Capability – I did not discuss as I was not sure (still not as I have not seen drawings yet) if the KPC (key product characteristics) have been defined on the drawings.
- EMD will provide the KPC's to TULOMSAS.

Technical Issues

- The final drawings of the under frame are handed over to TULOMSAS on CD format. In case of a revision to these drawings EMD will send the revised drawings to TULOMSAS. The list of drawings not included on the CD are outlined and given to EMD.
- Approximately 60% of the flat patterns are provided to TULOMSAS on soft format. A dimensioning is necessary to be performed on these drawings.
- Cable brackets were initially not included in non-checked drawings but this job is required to be accomplished by TULOMSAS. Some additional labor and material cost should be added.
- TULOMSAS asked for approval to use EDC 3244 steel quality instead of EDC 3241.
- Technical specifications of pipes are sent to EMD by e-mail.
- St44-2 steel quality is offered for Walkway plate material by TULOMSAS. (Related standards DIN59220, BS EN 10051, ASTM A-786)
- St44-2 steel quality is offered for U-Sections and Angels.

- Mr. Grzybowski has been informed about the TULOMSAS request related to the production changes of hand brake chain part, air duct flange and of the part with number 8084360.
- TULOMSAS asked for EMD approval to use diamond shape steel sheets for step floor.
- Design Change Process-Discussion on the design change requirements as they come up during the manufacturing process.
- In case any design change is needed, Requirement of Design Change form will be used. If EMD makes any changes in drawings, EMD will inform TULOMSAS about the changes. EMD will provide training on ECR/MCR process 2–3 weeks prior to manufacturing.

Localization Issues-Bogies

EMD and TULOMSAS have performed a corporate work for the manufacturing of DE 33000 type locomotives with a local content of 51 %. It has been decided to manufacture the bogies and the bolsters locally, so as to reach the 51% local content. Furthermore:

- As EMD, we have been supplying these bogies from the companies in US or foreign countries. This bogie is not a new design bogie. It has been assembled in thousands of locomotives and they are approved. We did not encounter any problems.
- All kinds of documents such as drawings, technical specifications had been given to TULOMSAS by EMD to be able to manufacture the bogies. TULOMSAS made ASÇELİK manufacture the bogies according to a/m documents provided by EMD. There will be no problem in case the bogie manufactured and tested according to the mentioned documents.
- The bogies manufactured in Turkey do not need to be tested by any additional inspection company/organization or do not need to be tested/tried for many kilometers.
- However, TULOMSAS has made the İTÜ (Istanbul Technical University) perform static and dynamic tests. The test report which the ITU prepared is appropriate.
- Besides, TULOMSAS has the capacity to perform metallurgical, chemical and dimensional tests.

Briefly, since these bogies which are manufactured and tested according to the EMD documents have been tested for many years, they do not need to be tested by a company or on the railroad any more.

Commercial Issues-Offset

- Regarding some steel castings, the specifics will be discussed and decided by both sides' material procurement coordinators.
- There were attempts in the past by EMD-GM's to have TÜLOMSAS as its supplier. As a result, it was promised to have TULOMSAS included on the list of distribution for these new RFQ's.
- In addition, it will be continued to review, and there will be other possibilities which the parties can include for future discussions.
- TULOMSAS will manufacture locomotive underframes for EMD as part of a large project in South Africa. This venture will continue to add new levels to the depth and synergy of our mutual cooperation.
- This is to certify and put in effect the reciprocal understanding of both parties for the manufacture and exportation of 50 units GT 26-MC locomotive underframe by TULOMSAS in Eskişehir Turkey.
- EMD will inform all the contact information of the shipper and the South African company.
- EMD will finish the opening of the Letter of Credit (L/C).
- EMD will provide information about how to load the under frames on to the ship

References

Documents of TULOMSAS, TCDD and TURKISH TREASURY (2002–2009) Ankara, Turkey

Kantarci M (2011) Speed Rail Projects in the World, Europe and Turkey, Okan Üniversitesi, 18 Feb 2011

Karacasulu N (2001) International Technology Transfer Process and Methods, Foreign Trade Magazine, Number: 20, Jan 2001

Ünal S, Kantarci M, Karaman S (2004) Railway systems improvement strategies, mimeo, 2004

UNIDO (1976) Guidelines for development of industrial technology in Asia and the Pacific, United Nations Centre-Bangkok

Chapter 12
Technology Transfer Through Joint Ventures in the Aviation MRO Industry: The Case of Turkish Technic

Fuat Oktay and Vehbi Özer

Introduction

Technology transfer from developed countries to developing countries is important for economic development and growth. Developing countries are using both national policies and international agreements to increase international technology transfer (ITT). National policies vary from the general, such as education and intellectual property rights (IPR) protection, to the specific, such as tax incentives for purchase of certain types of capital equipment (Hoekman et al. 2005).

In the current age of economic globalization, technology transfer has gained international dimensions. It has played a crucial role not only in the industrial growth of developing countries, but also in enhancing the competitiveness of their enterprises in the international market. Technology transfer between countries are through many different channels such as foreign direct investment (FDI), joint ventures, licensing, and subcontracting (Miyake 2005).

Joint ventures between companies in developing countries and foreign companies, among other technology transfer methods, have become a popular way for both managements to satisfy their respective objectives. They offer an opportunity for each partner to benefit significantly from the comparative advantages of the other. Local partners bring knowledge of the domestic market and local labor market. Additionally, local partner is familiar with government bureaucracies and regulations, which is valuable to the new venture. Foreign partners can bring advanced processes and product technologies, management know-how, and access to export markets. Joining with another company to form a new venture lowers capital requirements for both sides, relative to going it alone (Miller et al. 1997a, b).

F. Oktay (✉) • V. Özer
Turkish Airlines Technic Inc, Istanbul, Turkey
e-mail: fuatoktay@thy.com; vozer@thy.com

M.A. Yülek and T.K. Taylor (eds.), *Designing Public Procurement Policy in Developing Countries*, DOI 10.1007/978-1-4614-1442-1_12,

In this chapter, technology and technology transfer is discussed in section "Technology Transfer". In section "Strategic Method", the joint venture establishment process for technology transfer is analyzed, together with a proposed strategic method that can be used while establishing joint ventures, especially for the Maintenance, Repair, and Overhaul (MRO) industry in aviation. In section "The Case of Turkish Technic", the paper gives brief information about Turkey's aviation industry, MRO strategies, and Turkish Technic's technology transfer strategies based on its strategic plan. Benefits from these strategies, and the importance of technology transfer, are given as a conclusion.

Technology Transfer

Before giving the definition of technology transfer, it is important to define what technology is. One of the most basic and widely used definitions of technology is "information, know-how and experience which is necessary and used in order to produce a good or a service". Pier defines technology more broadly: "Technology is a body of knowledge, tools, techniques, and innovations derived from science and practical experience, which is used in the development, design, production, and application of products, processes, systems, and services" (Pier 1989).

Technology transfer is the process of sharing skills, knowledge, technologies, methods of manufacturing, among governments and other institutions. It ensures that scientific and technological developments are accessible to a wider range of users who can then further develop and exploit the technology into new products, processes, applications, materials, or services. It is closely related with knowledge transfer (Hargadon 2003), since whether from basic research to applied technology or from one firm to another, the transfer of technology is fundamentally a matter of the flow of human knowledge from one human being to another (Teece 1977).

One of the main drivers of technology transfer is to catch the leaders in the industry as it can be seen in Fig. 12.1 (Akyos and Durgut 2001). Technology can be produced by a company or a country using its own efforts to create new information, or by stimulating technology transfer strategies. Technology transfer helps to catch the leaders in industry and drives economic growth.

Technology Transfer Methods

Technology transfer from developed countries to developing countries can follow several channels. Yang categorizes these channels into four major categories (Yang 2007). First, technology holders can export the goods directly. Second, they can set up subsidiaries through FDI (Foreign Direct Investment) and control the production process. Third, they can choose to license technology to developing countries. Finally, they can form a joint venture with a host firm under a joint production

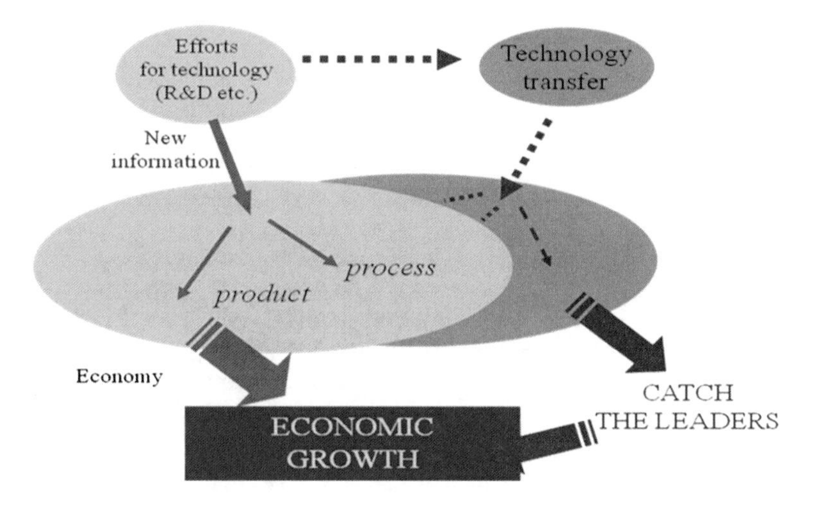

Fig. 12.1 Technology transfer. *Source*: (Akyos and Durgut 2001)

and technology-sharing agreement. Another study shows that movement of people among countries, associated with nationals having studied or worked abroad for a limited period, or the inward movement of foreign citizens, constitutes another potential channel for ITT. A good example of this is the recent experience of India developing a software industry and related services, with the development of the necessary skills within the country through nationals that have acquired those skills outside country. It illustrates that payoffs from such movements (outward in this case) may take time to materialize benefits for the country, although they can be large eventually (Hoekman et al. 2005).

Karacasulu classifies technology transfer methods into two main categories (Karacasulu 2001): namely Direct Technology Transfer and Indirect Technology Transfer methods. Karacasulu includes Joint Ventures as a subcategory of Foreign Direct Investment in this classification.

Direct Technology Transfer methods include the methods below:

- Foreign Direct Investment
  - Joint Ventures
  - Direct Investments
- Technology Transfer Agreements
  - License Agreements
  - Management Agreements
  - Technical Support Agreements
  - Turn-key Contracts
- Machine and Equipment
- Financial Leasing
- Employee Foreign Experts
- Free Zones

- Subcontractor
- R&D Activities

Indirect Technology Transfer methods include the following:

- Public Knowledge
- Education
- Human Resources

Joint Ventures as a Tool for Technology Transfer

A joint venture is a legal entity formed between two or more parties for the purpose of undertaking an economic activity together. Harrigan defines joint ventures as contractual cooperative forms through which two parent firms share ownership of a third organization, the joint venture (Harrigan 1988).

In the global economy, firms that wish to remain competitive in the international business environment realize that they need to expand into foreign markets. Thus, the use of international strategic alliances as a tool of competitive strategy becomes important for international business. International strategic alliances are important for the expansion of a firms's activites in international markets, as well as for sustaining the competitive position of the firm in the global business environment (Hajidimitriou and Georgiou 2002).

A specific type of international strategic alliance that has attracted a great degree of attention is the international joint venture (IJV). An IJV is described as a new business entity that is created by two or more legally distinct organizations (the parents), among which at least one is head-quartered outside the country where the new firm is located. Parent organizations hold ownership interests, and actively participate in the decision-making activities of the jointly owned business entity (Geringer 1991; Park and Ungson 1997). A driving force for the extensive use of IJVs is that the firms increasingly realize that the enormous costs and risks associated with international undertakings make it difficult for them to remain self-sufficient (Inkpen and Li 1999).

Although there may be various ways to gain new knowledge, the establishment of international joint ventures (IJV) is seen as one of the most efficient means to learn or absorb technology (Kandemir and Hult 2004), especially when the transfer occurs from developed to developing economies. Developing countries think that JVs help to set up their own industry which the developing countries may not be able to do by itself. As an example you can not establish an aircraft engine maintenance facility if you do not have expertise on this. Setting up JVs can help to gain this expertise. Some developing country governments impose restrictions on foreign ownership in order to protect fledgling local competitors from larger and established foreign competitors. China, as an example, does not allow joint ventures to have more than 49% foreign ownership. Therefore, if a multinational company wants to enter a developing country and produce locally, it must first find a local partner. Second, joint production and intensive training are appropriate for the transfer of some forms of knowledge. Hedlund distinguishes among three forms or knowledge: cognitive knowledge, skills, and knowledge embodied in products and services (Hedlund 1994). Cognitive knowledge and knowledge embodied in products are relatively easy to transfer, either through written documents or sales of machines. Meanwhile, skills, such as complex engineering processes, tend to be embedded in organizational routines and, therefore, are difficult to extract from another firm (Inkpen and Beamish 1997). The tacit aspect of skills can only be transferred through personal interaction and cooperation, which may be achieved more readily by joint ventures rather than licensing, FDI or goods trade (Yang 2007).

One final reason of why technology transfer through joint ventures is more effective than other channels is that effective technology transfer through learning would include three steps, namely transfer, transformation, and harvesting (Beamish and Berdrow 2003). Transfer is the migration of knowledge from technology holders to technology receivers, and traditional FDI focuses on this aspect only. Transformation is the potential creation of new knowledge through interpretation, cooperative production process, and interaction of individuals. Meanwhile, harvesting refers to the flow of transformed and newly created knowledge from joint ventures back to the partners' parent firms. In this way, joint ventures provide the possibility of transformation and harvesting, which facilitates more effective and profound technology transfer (Yang 2007).

Strategic Method

There are several studies and conceptual method proposals about joint ventures for specific areas, such as: partner selection (Hajidmitriou and Georgiou 2002; Luo 1998), control in joint ventures (Yan and Gray 2001; Tezölmez and Gökşen 2007), selecting management talents for joint ventures (Adobor 2004), and framework proposals for joint venture negotiations (Luo 1999). However, there is no study on a strategic method that will include all the steps for joint venture establishment that involves technology transfer.

Hajidmitriou and Georgiou (2002) defined the joint venture establishment process in four stages: partner selection, negotiations, agreement formulation, and operation and management of the IJV. This chapter will propose a more detailed method for joint venture establishment. (See Fig 12.2)

The steps of the method are as follows:

- Developing strategic plans

This step is the first and the most important one. Establishing joint ventures shall support the strategic plans of the company, therefore at the beginning the company should prepare its short and long-term strategic plans.

- Determining the strategies that would require technology transfer

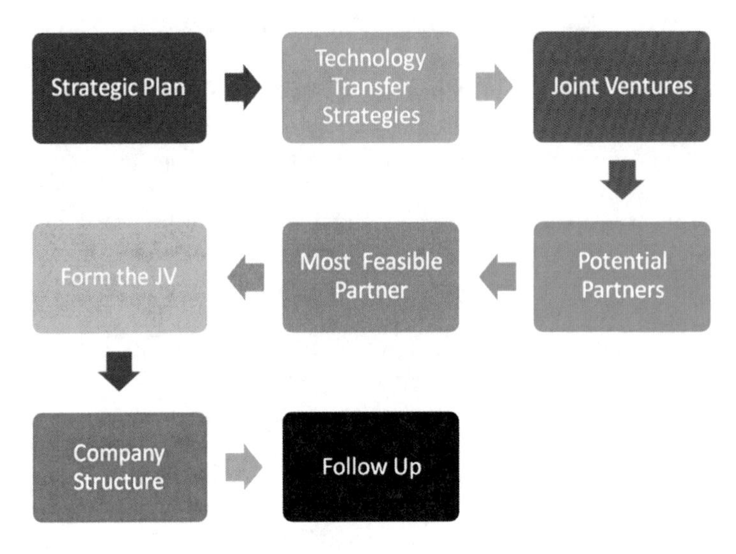

Fig. 12.2 Proposed strategic method

At this step, the company should identify the strategies that would require technology transfer. Strategies would necessitate technology transfer, if the company decides that it cannot reach its targets with internal sources and that technology transfer is crucial. The company should decide on which transfer channels shall be used in order to obtain these technologies.

- Determining technology transfers that would most efficiently be realized through joint ventures

Previous research showed that the true value of technology is related to applications used for specific purposes. Descriptions of procedures and operations, i.e. the codifying of knowledge to operate equipment, constitute an important aspect of technology transfer. As a result, "tacit" knowledge becomes much more accessible, and this, in turn, is important to the transfer of knowledge (technology) across organizations and borders. It should be noted, however, that the complete coding of knowledge is seldom possible, which suggests that learning-by-doing under supervision will often be an important element in the technology transfer (Gronhaug et al. 1999). Therefore, establishing joint ventures seems to be the most appropriate and effective way for technology transfer, especially in situations where the work is very complex and experience is an important asset At this step, the company should identify the strategies that can only be realized through establishing joint ventures.

- Determining the potential partners for joint ventures

The company should search for potential partners, and each potential partner should be included in the long list. In addtion, at this step, the company should also define its expectations from the joint venture, which can help the company bring down the long list to a short list. The company should also understand and define its "red lines", which are the minimum expectations and conditions under which such a joint venture can be formed. For example, if the company wants to have the majority control of the joint venture, this is a "red line" for the company, then the potential partners should be selected according to this condition. If the potential partners do not satisfy this condition, then they will be deleted from the list

- Selecting the best/most feasible or compatible partner

Selecting a partner that satisfies the conditions established in the "red lines" is important when establishing joint ventures. Partners must identify appropriate criteria, as well as decide on the relative importance of each criterion (Luo 1998). Broadly, the criteria can be classified into three categories related to (1) tasks or operations; (2) partnership or cooperation; and (3) cash flow or capital structure. Operation-related criteria are associated with the strategic attributes of partners, including marketing competence, relationship building, market position, industrial experience, strategic orientation, and corporate image. Cooperation-related criteria often mirror organizational attributes, such as organizational leadership, organizational rank, ownership type, learning ability, foreign experience, and human resource skills. Cash flow related criteria are generally represented by financial attributes exemplified by profitability, liquidity, leverage, and asset management. Even though a potential partner has high scores on strategic attributes, it may result in an unstable joint venture in case it has not strong organizational and financial characteristics (Luo 1998).

During the IJV formation stage, partners should undertake a formal analysis of the financial viability of the proposed venture, and undertake a due diligence analysis to ensure that what the partner has promised is actually delivered. During the negotitation stage, it is recommended that advice be obtained from third parties, particularly consultants who can provide an objective view on the proposed venture, in order to avoid problems from the beginning (Glister et al. 2003). A due diligence analysis is especially important at this step to confirm the benefits from the potential partner.

- Forming the Joint Venture
  - Strategic joint venture agreement
  - Other side agreements

Technology providers are interested in protecting their intellectual property. Therefore, they want to set limits on where and how the technology can be used by the joint venture and to place restrictions on who controls derivative technologies, no matter where developed. Important aspects in forming joint ventures may include, among other things, technologies not yet developed by either side. This needs to be covered in the agreement, including the terms under which they are to be made available to the venture. The developing country partners hope

Table 12.1 Technology transfer

Importance and difficulty of negotiating points in joint venture agreements (percentage of respondents noting category)

(% of respondents)	Important	Difficult
Equity structure	80	33
Technology transfer	78	26
Marketing issues	45	28
Staffing issues	44	26
Dividend policy	42	21

Source: Miller et al. (1997a, b)

to set bounds on the royalties and fees they will have to pay the providers, especially as the technology becomes older, and to broaden the joint venture's control over its use (Miller et al. 1997a, b).

Table 12.1 shows the result of a survey on the importance and difficulty of the negotiating points in joint venture agreements, covering responses from 75 joint ventures from 6 different countries. Based on the results, it is seen that technology transfer is one of the most important and difficult points to agree on during negotiations for joint venture agreements (Miller et al. 1997a, b).

- Forming the company that would serve to realize the business/partnership goals and that would best serve the technology transfer

Partners must clarify issues of management control, management structures, management roles, and how they will rotate. This ensures that there is no ambiguity, and that the managers know what their respective future roles will be. Data indicates that, through the formation process and the ongoing operation of the venture, it is important to identify which party is responsible for delivering the end result. The party must also know how the implementation plan will evolve as the international joint venture matures. It is also necessary to define clearly who is responsible for each activity and to avoid having more than one person responsible for the same thing. This prevents one party playing off the other. It also prevents a more senior manager from playing politics between the two. It should also be made clear exactly who is responsible for which set of decisions. This means identifying the levels of decision-making, which decisions go to the parents, the IJV board, and which ones are handled by the IJV managers, which requires effective communication between all the parties. It is also required that effort be put into team training in order to encourage practical working together on the part of both sides. Partners should recognize that it will take some time to bring together different systems, ways of thinking, approaches to problem-solving, and to resolve the personality clashes that are inevitably going to occur. A succession plan for senior management should be developed in order to determine who will operate at the strategic level (Glister et al. 2003).

The organization of the joint venture should be aligned with the technology transfer. There should be a good mix of expatriate and local managers in order to allow technology transfer and cultural change to occur (Miller et al. 1997a). A good idea would be to assign an expatriate to the head of a department, which is directly involved in technology transfer. This leads to increased technology transfer, which benefits the partner who will transfer the technology However, a local manager should support the expatriate for the absorption of the technology by a local firm, periodically following up the joint venture against the original strategic goals.

This follow up or review process is important since parties have stakes on joint venture outcomes. Furthermore, projections of this kind are based on certain assumptions, which may either be implicit or they may have been explicitly declared (Gronhaug et al. 1999). Such a review will help companies gain a better understanding of whether the expected outcomes are realized or if any corrective action needs to be affected.

The Case of Turkish Technic

Turkish Technic is the leading MRO provider in Turkey and is one of the leading MRO providers in its region with more than 2,500 employees. It has three maintenance hangars in Istanbul and one new maintenance hangar in Ankara. Turkish Technic provides aircraft maintenance for both Airbus and Boeing fleets, Auxiliary Power Unit (APU) and component repair and overhaul. Turkish Technic used to be a division of Turkish Airlines until May 2006. Later, it was separated from the Turkish Airlines and became an independent MRO industry with the purpose of serving not only Turkish Airlines but also other customers worldwide.

Besides being a leader in the MRO industry, one of the most important missions of Turkish Technic is to enhance Turkey's MRO capabilities in aviation and add a competitive advantage for Turkey.

Turkish Technic is within the top ten biggest MRO companies in terms of total revenues and revenue per employee (see Figs 12.3 and 12.4).

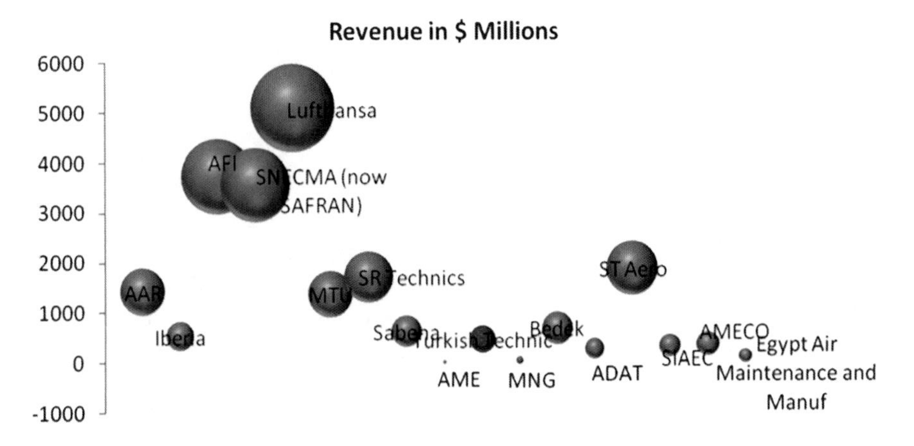

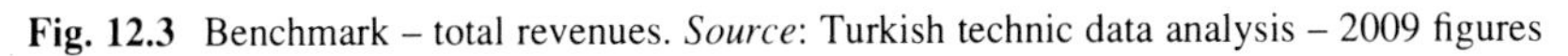
Fig. 12.3 Benchmark – total revenues. *Source*: Turkish technic data analysis – 2009 figures

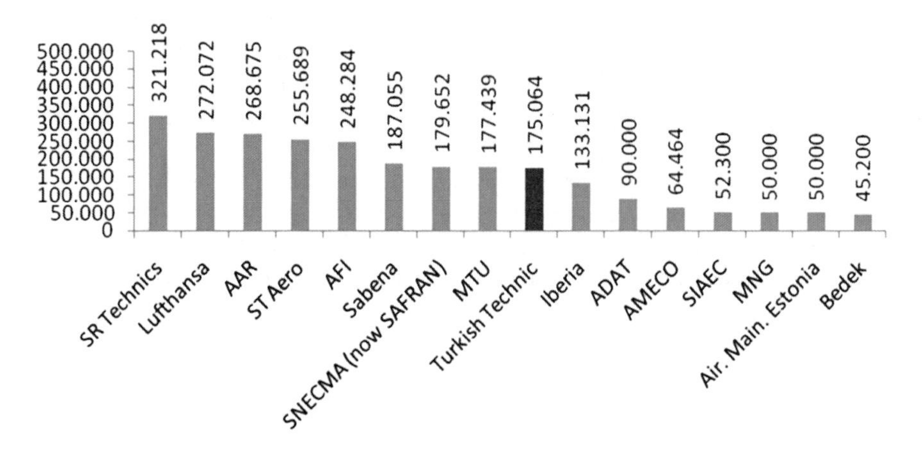

Fig. 12.4 Benchmark – revenue per employee. *Source*: Turkish Technic Data Analysis – 2009 figures

Turkish Aviation Strategies

The Ministry of Transport and Communication of the Republic of Turkey has created its Strategic Plan 2009–2013. Aligned with this strategic plan, the tenth Transportation Forum was held between September 27 and October 1, 2009, with the objective to create goals for the Turkish aviation industry, such as increasing the competitive advantage of Turkey in the aviation industry

As a result of this forum, there are four strategic road maps relating to aviation and MRO in aviation published as listed below:

- Locally manufacture at least two internationally known aircraft types, one/two turboprop and jet aircraft.
- Certificate an aircraft and/or aircraft parts by Turkey.
- Increase the local manufacturing content for aircraft body
  - Reach 20% for narrow body
  - Reach 8% for wide body
- Capture 10% of the Global MRO Market

Figure 12.5 shows that the 2009 MRO market was around USD45 billion. The 2023 Global MRO market is forecast to be more than USD70 billion. The 2009 MRO market capture rate from Turkey is less than 1.5%, based on the Turkish Technic data.

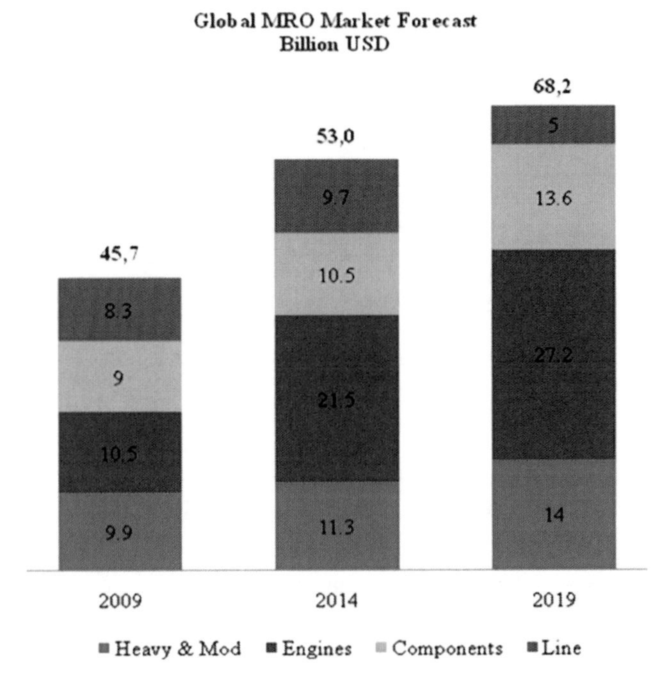

Fig. 12.5 Global Maintenance, Repair, and Overhaul (MRO) market forecast. *Source*: TEAMSAI 2009 figures

The current number of commercial aircrafts in Turkey is around 320. This number is forecast to reach 750 aircraft by the end of 2023[1] .The data show that Turkey's MRO Market will expand in later years (see Fig. 12.6).

Turkish Technic Strategies

As the industry leader, Turkish Technic has also created a Strategic Plan for 2010–2015, which aligns with the strategies of the Ministry of Transport and Communication. Three main objectives of this strategic plan are listed as follows:

1. To become a preferred top five MRO Center in the world
2. Increase and adopt Quality and Productivity with the IT-based approach
3. Build Environmental and Social Responsibility

[1] DHMI – General Directorate of State Airports Authority – figures and Turkish Technic Marketing Data.

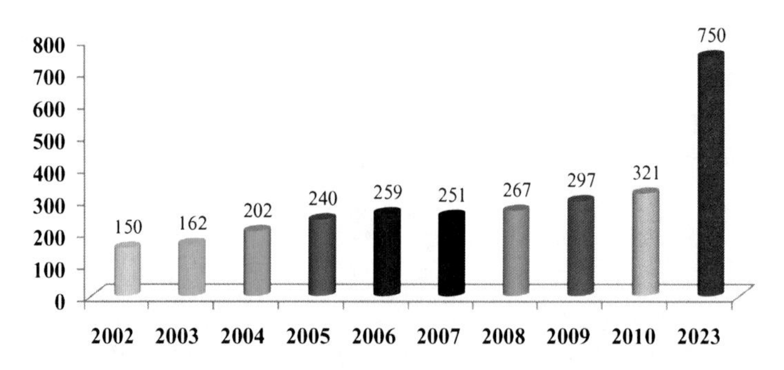

Fig. 12.6 Turkey Commercial Fleet Growth. *Source*: DHMI Data and Turkish Technic Forecast

Turkish Technic defines its strategic goals and strategies that require technology transfer in line with its vision to become a preferred top five MRO Center in the world. For this it aims:

1. To reach USD1.5 billion in 5 years with a customer oriented approach. For this it will have to:
   - Focus on component business
   - Become a "One-stop shop" until the end of 2011
   - Become a global and competitive player in the region, including the Middle East, North Africa, Middle Asia, and Eastern Europe, within 5 years
   - Position in the regions which will make Turkish Technic more competitive
2. To increase the share of other customers' revenues within the total revenues to at least 55%. For this, increase the product mix according to the customer needs.
3. To enter into new areas of business and get 2% of revenues from these areas. For this:
   (a) Enter into repair, manufacturing, and modification business segments
   (b) Become a risk sharing partner in global projects
   (c) Enter into military MRO business
   (d) Enter into private jet and helicopter MRO business

Joint Ventures as Technology Transfer Tools for Turkish Technic

In order to realize these strategies, Turkish Technic is planning to make Istanbul a significant MRO hub for aviation within the region, for which the company has started a new investment drive of around USD500 million at Sabiha Gökçen International Airport. The new international maintenance, repair, and overhaul

Fig. 12.7 HABOM project model

center project by Turkish Technic at the Sabiha Gökçen International Airport is named as the HABOM Project (Havacılık Bakım Onarım ve Modifikasyon Merkezi), which stands for Aviation Maintenance, Repair, and Overhaul Center.

The HABOM Project consists of a new facility designed as a center of excellence, including hangars, component back shops, training center, and other social buildings. The new hangars will hold twelve narrow-body and three wide-body aircraft, which will have almost doubled the existing capacity of the airport. The first phase of the facility will be in operation within 2011 (see Fig. 12.7). By the year 2020, HABOM is estimated to generate a USD one billion share from airframe and component maintenance segments.

The HABOM Project's history goes back to the ITEP Project of Turkey. The goal of the ITEP Project was to build an advanced technology and industrial zone in the Pendik area of Istanbul. It consisted of an airport, MRO center, Techno park, and a Trade and Commercial Area for exhibitions and hotels. The first phase of the project was completed by the construction of the Sabiha Gökçen International Airport in 2001.

Turkish Technic started to look for opportunities to establish an MRO center at the Sabiha Gökçen International Airport since 2004. In this connection, a joint venture was considered to enable technology transfer in the MRO industry by involving global players. A good example in this connection is that of Lufthansa Technic, one of the major players in the MRO industry, which is using partnerships and joint ventures to increase it global presence and network. Lufthansa Technic's facilities extend from North America to India, and from Europe to China. By establishing joint ventures with global players, Turkish Technic is showing its readiness to position

Fig. 12.8 Picture from TEC

itself as a global player and benefit from the technology transfer to be realized from prospective international partners.

In addition, joint ventures constitute a tool for technology transfer. Teece (1980) notes that the transfer of know-how involves a strong element of learning-by-doing. This is of crucial importance for the choice of the exchange mode. When it is difficult to determine the separate elements to be transferred, then close contact and interaction between the parties involved will often be neccessary in order to achieve the intended outcome of the transfer. A Joint venture enables this more easily, as compared to other methods, because it naturally allows for greater interaction among the people involved in the whole process.

Turkish Technic's first joint venture is Turkish Engine Center (TEC) (Fig. 12.8), which is a joint venture with the global engine manufacturer Pratt & Whitney. Turkish Engine Center has started its operations in early 2010 with the goal to become a global player in engine MRO. With this joint venture initiative, Turkish Technic aims to increase its engine maintenance capacity, as well as its engine parts repair capabilities. The main form of technology transfer of TEC is increasing repair capabilities, and finding the best practices for engine overhaul within the Pratt & Whitney global network. As TEC increases its repair capabilities, more work will be done in Turkey, rather then sending parts to other countries for repairs.

Turkish Technic's second joint venture is with Goodrich Aerostructures, Inc. to provide service for nacelle and thrust reverser parts. The joint venture agreement was signed in 2010, and the joint venture is going to be in operation in 2011, with which Turkish Technic will gain thrust reverser and nacelle overhaul and repair capability. Therefore, this is also a good example for technology transfer.

Turkish Technic, as a part of its technology transfer strategy of developing local manufacturing and repair capabilities, announced recently the establishment of the

Turkish Cabin Interior Center as a joint venture with a domestic company. It will focus on the interior work for the aircraft, extending pool and repair services, and seeking new joint ventures in other geographic areas outside of Turkey in order to access global markets.

Technology Transfer through Joint Ventures

Technology Transfer Through TEC (Turkish Engine Center)

As a result of the joint venture, TEC and Turkish Technic would have realized the technology transfer through the items below:

- The best practices from Pratt & Whitney's other engine centers

The Pratt & Whitney network has seven engine centers worldwide, including TEC. At times, the engine center workers and the executives from all seven locations come together to share their best practices which can be used in engine overhaul. This creates synergy and an opportunity for transfer of technology and know-how. Best practices help TEC to improve its processes. For example, Turkish Technic has more experience on overhauling the engines manufactured by the CFM International, while Pratt & Whitney has more experience on overhauling V2500 type engines. When these experiences come together and best practices are shared, TEC benefits from all this know-how.

- Achieving Competitive Excellence (ACE) for continuous improvement

ACE is Pratt & Whitney's operating system for continuous improvement. TEC has also started to use this operating system, as yet another good example of technology transfer. ACE is recognized to provide best-class quality in its products and services with a focus on increasing efficiency and reducing waste. Turkish Technic has also been implementing various lean manufacturing and management techniques for continuous improvement. As a result, employees of TEC create a competitive advantage by easily adopting to work with the ACE operating system. All ACE employees, including managers and executives, have been trained to use the ACE operating system efficiently.

- Benchmarking

Benchmarking is the process of measuring an organization's internal processes, then identifying, understanding, and adapting outstanding practices from other organizations considered to be the best-in-class. The Pratt & Whitney network, which consists of seven engine centers worldwide, including TEC, creates a good opportunity for the engine center managers and executives to reach different data sources to use for benchmarking. Benchmarking also helps managers and executives to improve their processes and give targets for their key performance indicators.

- Engine Shop Layout

To create a state of the art shop, the layout of the shop has been designed with a benchmark of other engine centers. Visual factory concepts are used to create a lean shop, which allows expansion in later years without making significant changes in the layout. An efficient shop layout is created, when the experience of workers from Turkish Technic are transferred to TEC, combined with the know-how of workers from Pratt & Whitney. The layout of an engine shop is very important and it has direct effects on engine overhaul. A good layout helps to improve processes, reduces waste, and decreases the time required to overhaul an engine.

- Facility Design

TEC's facility is designed to satisfy LEED Gold requirements. LEED stands for Leadership in Energy and Environmental Design. The United States Green Building Council (USGBC) created LEED as a rating system for green building. Green building refers to the design, construction, and operation of buildings in an environmentally friendly way. LEED promotes a whole-building approach to sustainability by recognizing performance in five key areas of human and environmental health: sustainable site development, water savings, energy efficiency, materials selection, and indoor environmental quality. TEC's LEED Gold certification was awarded in July 29, 2010. TEC is the first aviation-related facility and second building with LEED Gold certification in Turkey. Some of TEC's sustainability features within the facility are:

  - High performance energy efficient windows providing daylight to the facility, reducing electricity requirements and reducing the need for additional cooling.
  - Added to large windows, solar tubes throughout the whole facility introduce daylight to 100% of the regularly occupied spaces.
  - More than 10% of the total materials used come from recycled sources, reducing the need for raw material extraction.
  - Rainwater is being harvested into 500 m^3 tanks from a large roof area. It is then filtered to municipality domestic water standards and is used throughout the facility, as well as in landscape irrigation. Overall, water savings exceed 60%.

- Additional Repair Capabilities

After the formation of TEC, a subcommittee was formed that included members from Turkish Technic, Pratt & Whitney, and TEC in order to identify additional repair capabilities that TEC can eventually have. The first phase of the work has been completed to identify these repair capabilities. When TEC eventually attains these capabilities, there will no longer be any need to send the engine parts outside the country for repairs.

- Core technology

Turkish Airlines Technic has accumulated significant experience and expertise during its seventy-seven year history in the MRO industry, while Pratt & Whitney is one of the major global aircraft engine manufacturers in the World. Combining the strenghts of these two companies will facilitate technology transfer relating to the core technology of an aircraft engine.

- Training of employees

Technicians are routinely sent to engine centers in other countries for training. As it was mentioned in section "Introduction", the technology transfer is mostly related with human beings and tacit knowledge, which can only be transferred by doing actual work. Training gives an opportunity for such interactions and helps technology transfer.

- Information Technology

TEC is using the *Systeme, Anwendungen, Produkte*: "Systems, Applications and Products" (SAP) system, which has been used in Pratt & Whitney engine centers for a long time and had been customized based on the engine center requirements. This is the first time that this customized and developed SAP system has been implemented by a Turkish company. Therefore, it is localized as well based on local financial regulations. This constitutes a valuable experience for both parties, since localization adds value to the SAP system and international customizations help TEC use it efficiently.

Conclusion

Turkey hopes to reach the target of 10% share in the world MRO market by 2023, which will generate more than USD7 billion in revenues. However, Turkey's existing MRO capabilities will not be sufficient to reach this target without gaining new capabilities and focusing on products and services with more complex technological content. Technology transfer is, thus, crucial as one of the drivers to help Turkey reach her strategic goals in the MRO industry in aviation.

Furthermore, although technology transfer can be managed by different mechanisms, joint ventures with international companies are an efficient way for sustainable technology transfer. This is because (1) some MRO activities are complex and technology-driven, (2) skills, experience, best practices, know-how, and learning-by-doing are important drivers of success in the MRO sector and an international partner can assist in these respects (3) an international partner can also help in accessing the global markets.

It should be noted that, although joint ventures have attractive features, many joint ventures fail (Killing 1982). The risk of failure can be reduced with a solid strategic map that is implemented carefully.

References

Adobor H (2004) Selecting management talent for joint ventures: a suggested framework. Hum Resource Manag Rev 14:161–178
Akyos M, Durgut M (2001) International technology transfer/local technologic talent, competition until where? how? panel, Gazi University, Ankara, 7–8 Nov 2001.
Beamish PW, Berdrow I (2003) Learning from international joint ventures - the unintended outcome. Long Range Plann 36:285–303
Geringer JM (1991) Strategic determinants of partner selection criteria in international joint ventures. J Int Bus Stud 22(1):41–62
Glister K, Husan R, Buckley PJ (2003) Learning to manage international joint ventures. Int Bus Rev 12:83–108
Gronhaug K, Hauschildt J, Priefer S (1999) Technology transfer through international joint ventures: the case of gamma. Scand J Manag 15:307–320
Hajidmitriou YA, GEORGIOU AC (2002) A goal programming model for partner selection decisions in international joint ventures. Eur J Oper Res 13:649–662
Hargadon A (2003) How breakthroughs happen: the surprising truth about how companies innovate, 1st edn. Harvard Business School Press, Boston
Harrigan KR (1988) Joint ventures and competitive strategy. Strat Manag J 9:141–158
Hedlund G (1994) A model of knowledge management and the N-form corporation. Strat Manag J 15:73–90
Hoekman BM, Maskus KE, Saggi K (2005) Transfer of technology to developing countries: unilateral and multilateral policy options. World Dev 33(10):1587–1602
Inkpen AC, Beamish PW (1997) Knowledge, bargaining power and international joint venture stability. Acad Manage Rev 22:177–202
Inkpen AC, Li KQ (1999) Joint venture formation: planning and knowledge – gathering for success. Organ Dyn 27(4):33–47
Kandemir D, Hult GTM (2004) A conceptualisation of an organizational learning culture in international joint ventures. Ind Market Manag 34(5):430–39
Karacasulu N (2001) International technology transfer process and methods. Foreign trade magazine, Number 20, January 2001.
Killing JP (1982), How to make a global joint venture work? Harvard business review, 1982, vol. 60 (May/June 1982), pp. 120–27.
Luo Y (1998) Joint venture success in china: how should we select a good partner? J World Bus 33(2)
Luo Y (1999) Toward a conceptual framework of international joint venture negotiations. J Int Manag 5:141–165
Miller R, Glen J, Jaspersen F, Karmokolias Y (1997), International joint ventures in developing countries, IFC Discussion Paper No. 29 Washington, World Bank, 1996.
Miller R, Glen J, Jaspersen F, Karmokolias Y (1997b) International joint ventures in developing countries. Finance & Dev 34:26–29
Miyake T (2005) International technology transfer, CACCI J vol 2
Park SH, Ungson GR (1997) The effect of national culture, organizational complementarity, and economic motivation on joint venture dissolution. Acad Manage J 40(2):279–307
Pier A (1989) Technology: a key strategic resource. Manage Rev 78(2):37–41
Polanyi M (1967) The growth of science in society. Minerva 5(4):533–545
Teece DJ (1977) Technology transfer by multinational firms: the resource cost of transferring technological know-how. Econ J 87:242–261
Teece DJ (1980) Economics of scope and the scope of enterprises. J Econ Behav Organ 1:223–247
Tezölmez SHU, Gökşen NNS (2007) Control and performance in international joint ventures in turkey. Eur Manag J 25(5):384–394
Yan A, Gray B (2001) Negotiating control and achieving performance in international joint ventures: a conceptual model. J Int Manag 7:295–315
Yang B (2007) Autocracy, democracy, and FDI inflows to the developing countries. Int Econ J 21(3):419–439

Chapter 13
Procurement Policy and Technological Development in the Turkish Aviation Industry: The Offset Experience of Kale Aero

Yalçın Yılmazkaya

Introduction

Kale Corporation is one of the largest industrial conglomerates in Turkey with yearly sales of approximately USD 1 Billion. Established in 1957 by one of the pioneering industrialists of Turkey, Dr. Ibrahim Bodur, KALE has its main business in the production of ceramic tiles (largest factory in the world under one roof) and mining, along with its technology and services group having several very successful technology-oriented companies in key sectors. The latter include Software and Consultancy (Kale Data), Defense (Kalekalip) & Aerospace (Kale Aero), (Kale–Baykar) and (Kale Pratt & Whitney), Gas Equipment (Kale Energy), and Construction Chemicals (Kalekim).

Today, Kale Aero, being a direct supplier for many world-known defense and aerospace companies such as Lockheed Martin, Boeing, and Airbus, has been recognized as one of the leading suppliers in many important international programs, including a critical part manufacturer for the largest defense program Joint Strike Fighter (F – 35) project. On the other hand, Kale–Baykar is a successful Joint Venture with its indigenous UAV systems having their mini UAVs under serial production and mission tested. New tactical UAV systems are also under development. Kalekalip however, being the core of the Defense and Aerospace innovation and know-how in the Group, has many products designed and manufactured for the use of the Turkish Armed Forces, including Automatic Grenade Launchers, Rocket Launchers, and Missile parts and systems.

Currently, Kalekalip is working heavily on the design and manufacturing of the indigenous infantry rifle for the Turkish Army. Kale Pratt & Whitney Uçak Motor Sanayi A.Ş., is a joint venture with Pratt & Whitney, a United Technologies Corp. (NYSE: UTX) company. The partnership will be operating in a new facility

Y. Yılmazkaya (✉)
Kale Group Companies Inc, Istanbul, Turkey
e-mail: y.yilmazkaya@servicesby.com

M.A. Yülek and T.K. Taylor (eds.), *Designing Public Procurement Policy in Developing Countries*, DOI 10.1007/978-1-4614-1442-1_13,

to be located in Izmir ESBAS Free Trade Zone. The product range of Kale Pratt & Whitney Uçak Motor Sanayi A.Ş., which will likely manufacture F135 parts, will not be limited to the F135 products only. The company is expected to expand to other commercial and military Pratt & Whitney product lines as well.

From Mold-Making to Defense and Aerospace

Corporate Venturing has been a well-implemented concept in the Kale Group. Kalekalip was established in 1969 as a molds and dies manufacturing shop within the Kale Group to support the main ceramic tile production. Die- and mold-making is a precision business and, over the years, Kalekalip has accumulated the technology and know-how to manufacture precision parts and systems.

In the mid-1990s, one of the largest defense contractors of the US, Lockheed Martin Missiles and Fire Control (LMMFC), a division of the Lockheed Martin Corporation specialized in rocket and missile production, was looking for companies in Turkey to help them in fulfilling their offset obligation due to a military sale to the Turkish Government. In this program, they were planning to put part orders to Turkish suppliers.

Defense and aerospace parts manufacturing is a completely new business with very high quality and delivery expectations along with stringent tolerances and a need for very special processes, tools, and special equipment.

The Kale Group had a vision for growth in technology-oriented sectors where the usage of high technology combined with a skilled workforce was high, but also the return on investment was much higher than in conventional businesses. Thus, the Kale management, believing their strength in creating a skilled workforce and with their already existing know-how in precision parts production capabilities, has decided to start a dialog with the LMMFC executives to understand their needs and define a way to collaborate with them in this offset program.

The Start with a USD 50,000 Offset Order

The needs and expectations were highly challenging. Furthermore, in keeping with the agreements, the defense contractors normally insist that the "foreign" companies meet certain contract standards for price, quality, and delivery to become qualified subcontractors. Moreover, the foreign companies are expected to meet the same contract standards as those used to select the qualified home country subcontractors. In many cases, this is a good business since the foreign subcontractor may end up becoming a longer-term supplier to the U.S. contractor of that particular item being produced. Yet, the risks are also very high.

Another challenge lay ahead while working with the large contractors like LM who has offset obligations globally. In most cases, sourcing organizations do not share the same goals and objectives with the offset program managers. Sourcing people tend to work with the suppliers within the vicinity of their production facilities and they are mostly happy with their performance in quality, delivery, and pricing. They also try to keep the risk to a minimum by doing so. This is because working with international suppliers sometimes creates delivery and quality issues. Thus, it becomes a big challenge to convince the latter to pass on new work to a new supplier in a far away land where they have limited control and with no proven record of quality and delivery. Especially for parts to be used in critical defense products and systems, where the national security is involved, the US buyers have always been and still are very reluctant to put an order to companies overseas. This was also the case for Kalekalip.

After a rigorous period of supplier prequalification by the LMMFC technical experts, negotiations have started. Although the LMMFC offset program leaders have developed a trust for Kalekalip's capabilities, convincing the sourcing organization in the LMMFC still lay ahead as a major challenge.

Kalekalip and the LMMFC program people have decided to face the challenges and moved forward. They have identified a critical part and Kalekalip decided to produce it in their facility competing with the existing US supplier's offer. A small trial order was placed for a total of USD 50,000 per annum. In order to produce the said part, new investment on machinery and equipment was required. Although the planned investment was so much higher than the order placed, The Kalekalip management has decided to make the necessary investment by taking this risk.

The sample part was produced as planned and has successfully gone through the qualification process. Subsequently, a long-term cooperation with the LM has started with this small step. The cooperation process was successful since

- Kalekalip supported LMMFC in the fulfillment of offset obligations in Turkey via local parts manufacturing, while also maintaining a good relationship with the local offset authorities in the country.
- This partnership has always been valued by the parties, as well as by the local offset authorities, since it brought on valuable projects in Turkey by creating jobs, investments, technology and know-how, and several other value-added programs for the benefit of the Turkish economy.
- In recognition of this valuable contribution, the Kale Aero-LMMFC partnership in offset fulfillment cooperation was chosen by the SSM (Turkish Offset Authority) as "The Best Offset Implementation Project" in Turkey.

Over the years, Kale became a star supplier for the LM Corporation and the monetary value of the cooperation has reached several hundred million USD. Today The Kale Aero is a major supplier to the Lockheed Martin Aero in the major JSF (Joint Strike Fighter) Project as the manufacturer of more than 200 highly complex parts.

New Horizons with the LMMFC

The ongoing collaboration with the LMMFC for Offset Program Support in other counties has added another dimension to this relationship. The Kale Group is continuously investigating new business opportunities that may be used as offset projects in countries where Kale has good relationships and knowledge about. This is expected to create a new win-win situation for both parties. Key points of this new collaboration are as follows:

- The LMMFC's view of the Kale Group as a strong and reliable partner
- The Kale Group's knowledge on Offset Fulfillment Programs
- Familiarity of Kale with the culture of the country
- The Kale Group's ability and capability to pursue new profit generating business opportunities

Key Elements for Success

There are certain basic but very important rules for success in such relationships. Many are obvious, but they are sometimes taken for granted. However, long-term success lies in the application of these simple rules:

1. Understanding the needs – CUSTOMER CTQs
2. Understanding the rules and regulations
3. How to perceive offsets – Obligation versus Opportunity
4. Ensuring two-way communication
5. Speaking the same language
6. Building mutual trust
7. Being open – sharing what can and cannot be done
8. Win-win

Understanding the Needs – Customer CTQs

Critical to Quality (CTQ) is a Six Sigma Terminology. Six Sigma is a statistical method to understand and overcome problems and defects in processes. The Kale Group applies this methodology in their production sites as well as in their relationship with their partners. Understanding the needs and expectations of one's customers, buyers or partners is the starting point for success. Misperceptions or assumptions usually end up with results satisfying none of the parties, hence jeopardizing the relationship. If one can understand the needs correctly, then one can set up the correct strategy to fulfill the needs. This also helps partners in using their limited resources correctly. Not more, not less since if one understands the needs well, one can plan well.

Understanding the Rules and Regulations

In the world of offsets, rules and regulations differ from country to country. In most cases, the offset guidelines are in the local language and direct translation may not provide the message that is expected from the offset authorities of that country. In this respect, a local partner may help with the correct understanding of the rules which helps the obligor in the success of the offset programs. On the other hand, there are international rules and regulations that are prominent and provide guidelines for transactions in defense-related articles (e.g., ITAR–International Traffic in Arms Regulations). In relationships with international customers, buyers, and partners, these rules need to be well understood, respected, and followed by the local companies.

How to Perceive Offsets – Obligation Versus Opportunity

In many cases, offsets are perceived as a burden by the primes. From the local industry perspective, however, offsets are also seen as a livelihood for survival. From this perspective, projects started with these objectives are short-lived. Primes look for shortcuts to fulfill their obligations and local companies want to make a living on offsets, which is also dependent on the duration of the offset program. Once the offset program is over, the relationship also vanishes. The value of offset support lies in the following:

- Helping to create offset project opportunities in the foreign country on behalf of prime
- Supporting the prime in handling offset transactions with the local authorities
- Building long-term relationships and future opportunities for both parties.

In this respect, the Kale Group and the LMMFC have looked at offsets as a first step of introduction and as an opportunity for a long-term mutually profitable relationship. The LMMFC valued Kale as a strong partner in Turkey and in the region, looking beyond the limited period of an offset deal and again beyond the limitation of a supplier–buyer relationship. The Kale Group also highly valued this relationship and cherishes the LMMFC and the LM Companies as a highly valued long-term partner.

Ensuring the Two-Way Communication

Communication is the key for success. However, true success lies in building the two-way communication between partners. Listening is an art. Information exchange is also a valuable tool for the success of all the strategies.

Speaking the Same Language

Since two-way communication is a key for success, partners should be able to avoid any misunderstandings between them by creating a common language in their relationships. In most cases, it is commonly seen that assumptions lead the way to most of the failures. It is a good practice to overcome the cultural differences in business transactions by creating standards and making sure that both parties share the same goals and objectives.

Building Mutual Trust

Mutual trust is the basis for long-term relationships. Yet it does not come overnight. Trust is built by making sure that both parties stick to their promises and make sure that they do not have any hidden agendas when working with each other. Partners should not forget that trust is built over time and once it is damaged, it takes a longer time to create the same environment between partners.

Being Open: Sharing What Can and Cannot Be Done

In some cultures, it is sometimes rude to say "no" to a partner upon a request although the chances of completing on time, on budget, etc. are highly at risk. This is done with no bad intentions. However, in many other cultures, this is accepted as a "job will be done" message where no contingency plans are made. This cultural difference is sometimes defined as "delinquency" and creates mistrust. It is strongly recommended that partners be open to one another and share the information upfront on any unforeseen risks with their partners. This way both parties can plan accordingly, and there will be no surprises that may jeopardize the trust between the parties.

Win-Win

One of the key elements of success is to understand that in creating a long-term relationship, there is no possibility that one party may win and the other loses. A strategy based on this kind of understanding may continue for a while until the other party starts looking for other alternatives. During the partnership, it is also good practice to discuss the objectives that are set in the beginning from time to time and make sure the current strategy to achieve these objectives are mutually accepted by the parties. Again, taking the partnership for granted may lead to short-term blindness and may jeopardize the partnership.

Other Global Customers and Partners of Kale

Today aside from the long-term successful cooperation with the LM, Kale is also a major supplier for many defense and aerospace companies including Boeing, Northrop Grumman, Pratt & Whitney, and Spirit Aerosystems.

The supplier–buyer relationship has recently led the way to a strategic partnership with Pratt & Whitney. Thus, the new Joint Venture in Turkey is called the Kale Pratt & Whitney Uçak Motor Sanayi A.Ş., a joint venture of KALE with Pratt & Whitney, a United Technologies Corp. (NYSE: UTX) company and a leader in the aircraft propulsion industry. The approximate investment will amount to USD 60 million. Kale will own 51% and Pratt & Whitney 49% of the new venture company. The Joint Venture will manufacture critical aircraft engine parts mainly for the F 135 (for JSF F35 A/C). The venture is planning to employ 500 people in the next 5 years.

Conclusions

Designing Efficient Public Procurement Policies to Foster Technology Transfer and Development Capacity in Emerging Markets is an important objective for countries in need of accessing new technologies and new markets. There are good examples in the world where offsets can be used to fulfill this objective. Provided that offsets are perceived as an enabler and a facilitator for creating mutually profitable relationships; it is proven that long-term and sustainable success can be achieved. However, local parties need to understand that it needs dedication and hard work to create the environment to absorb the technology that is shared and countries should create incentives for the technology providing party by accepting the fact that creating new technologies has a cost. The Kale Group and the Kale Aero have shown this dedication, trust, and patience where the partnership has grown from a simple supplier – buyer relationship under an offset program starting a USD 50,000 order to a much wider partnership covering larger regional opportunities.

Index

M.A. Yülek and T.K. Taylor (eds.), *Designing Public Procurement Policy in Developing Countries*, DOI 10.1007/978-1-4614-1442-1,

J